A Geographical Guide to the Real and the Good

A Geographical Guide to the Real and the Good

Robert Sack

ROUTLEDGE
NEW YORK AND LONDON

Published in 2003 by
Routledge
29 West 35th Street
New York, NY 10001
www.routledge-ny.com

Published in Great Britain by
Routledge
11 New Fetter Lane
London EC4P 4EE
www.routledge.co.uk

10 9 8 7 6 5 4 3 2 1

Library of Congress Cataloging-in-Publication Data
Sack, Robert David.
    A geographical guide to the real and the good / Robert Sack.
        p. cm.
Includes bibliographical references (p.   ).
    ISBN 0-415-94484-8 (HB : alk. paper) — ISBN 0-415-94485-6 (PB : alk.
paper)
    1. Geographical perception. 2. Human territoriality. I. Title.
    G71.5.S227 2003
    304.2′3—dc21

                                                                    2003001962

For
Fred and Barbara Lukermann

CONTENTS

PREFACE

I have written this book to explain why I think geography can help clarify moral issues. But it is up to the reader to decide if this is so. Do the arguments I make—that it is good to create places that increase our awareness of reality and increase the variety and complexity of that reality—help us understand what is morally better and what we ought to do?

I believe clarification is about all that one can accomplish in most philosophical arguments, and especially one on morality. This is why philosophers continue to argue, and why the illumination provided by good arguments clarify and can even inspire, but cannot resolve the important question. A good argument deserves consideration, but it is not the last word.

I offer the argument developed here in this spirit. It is a geophilosophical theory, written by someone interested in the relationship between geography and philosophy, an interest that I have followed in other books. I offer the theory with seriousness and conviction, and argue for it forcefully, but I recognize that is only a theory. I will remind the reader of this more than once, but to repeat it constantly prevents this or any other argument from developing. And while I believe it is important to recognize the limits of moral arguments, I also believe it is equally important to offer them, for they can clarify why some things are good and others not; if we do not try, then our silence suggests that good or bad is simply a relative matter, or of no matter at all.

The theory presented in this book draws on the work of many philosophers and geographers. I have tried to acknowledge my debts in the notes. I am especially conscious of those who I have explicitly asked for help. Here I want to acknowledge the invaluable assistance of members of my seminar, Better Worlds, and to John Agnew, Stephen Aschheim, Roger Bolton, Kim Coulter, Nick Entrikin, Robert Ostergren, Karen Sack, and Yi-Fu Tuan for reading (all, or parts of) the manuscript. They have helped me with the logic and the prose. Still, any mistakes and confusions in the argument are mine. Thanks also to Caitlin Doran for the illustrations, to Jonathan Durr for proofing the manuscript and page proofs of the book, and to David McBride and George Thompson for their help and encouragement. I especially want to thank Katya and Priscilla for offering me numerous and accessible examples of what it means to be good.

Introduction

1

GEOGRAPHY, THE REAL, AND THE GOOD

THE REAL AND THE GOOD ARE intricate and difficult concepts on their own. What possible connection do they have to each other and to geography, and what does it matter? This book explains how the three are intimately interrelated and how a geographic framework illuminates important qualities of reality and morality that may otherwise be overlooked. The connection among the three is underscored by the title of the book: *A Geographical Guide to the Real and the Good.*

Absent from the title is evil. Good and evil are linked, and this book addresses both. In fact, its most detailed examples are about evil. Why, then, does evil not appear in the title? The reason will become apparent from arguments I will present, and runs along these lines: Our evil acts stem from a lack of awareness. As a lack, evil cannot even be articulated or known for what it is without a conception of the good. Articulating the good as positively as possible provides a compass bearing that can guide us to do better and help us judge what is lacking—what is evil. This is why the title includes the good and suggests we need a guide to it, and does not give evil equal billing. But what about the geographical and the real, and their connections to the good? To set the stage, I will first introduce the role of the geographic.

A long-standing contribution of the discipline is that geography provides insight into how and why humans transform the earth. It reveals that our earth-transforming and landscape-creating capacities are basic qualities of our human nature: we are geographic beings. Earth-transforming refers to our changes to the natural environment, our transformation of the natural into a largely cultural environment,

3

and to the changes we make to that cultural world. For simple societies these alterations may be slight and go largely unnoticed by its members, as when a stand of trees becomes altered because it is used as a sacred grove, or when part of the forest is burned for planting and then that largely cultural landscape is allowed to return again to a forest, but one that is not exactly the same. For complex societies the changes become larger-scale, more rapid, and evident, as when entire forests are cleared to create vast farms and ranches, when fields give way to cities, and when these cities in turn are hollowed out by the creation of new suburbs. Whether in simple or complex societies, we transform nature into culture and then transform culture, and so on. These transformed natural and cultural environments constitute the reality that surrounds us.

Must we transform our environment? Geography's answer is yes. The root of the answer lies in the *geographical problematic,* which expands the idea about our geographical nature and can be expressed this way: *We humans are incapable of accepting reality as it is, and so create places to transform reality according to the ideas and images of what we think reality ought to be.*[1] This problematic provides the key to how geography, the real, and the good are connected. Unpacking the implications of this problematic is the task before us and our geographical place-making provides the key.

Place-making is essential to both our transformations and to our ideas and images of what reality ought to be. Place has a particular meaning in this book. It does not refer simply to a location in space. Rather it means an area of space that we bound and to some degree control with rules about what can and cannot take place. Place can be any size, from the small-scale of a room or a sacred grove, to the larger scale of a farm or a city, to a vast territorial unit such as a nation-state or empire; and some of the smaller ones can even move, as in the case of a railroad car, an automobile, a ship, or a plane. We as humans construct place because it is an essential tool that enables us to undertake virtually any and all projects. The projects of education, child rearing, work, community building, recreation, government, and practically everything else require places: schools, homes, factories and farms, neighborhoods, parks, cities and states. As a tool, place is as pervasive and important as language: we are place-makers and users, as we are language-makers and users.[2]

In not accepting reality as it is, we transform it through place-making. The transformations may be large-scale and continuous, or small-scale and infrequent. In place-making, we may create something new, or return to something that once was. Even if we want to keep things as they are, we still transform as we remove things that do not belong and prevent things from entering. In all of these cases, the reason we are manipulating the environment (from the sacred grove, to the room, to the globe) is to have reality become what we think it ought to be. The last part, "what we think reality ought to be," not only animates the entire process, but also provides the means of justifying or rejecting what takes place. True, the "ought" stems from needs, desires, and self-interest. But when pressed to justify these, we eventually realize that a fuller understanding of what "ought to be" embodies a moral idea based on a conception of what we think is good, and this idea of the good undergirds moral evaluations and concerns.

Geography is part of these evaluations. Places are essential even in our imagined worlds, and are intimately connected to the moral and what it means to be and do good. Imagining a project is to imagine it taking place, and in our imaginations, as in our actual practices, the basic act of clearing a place to create something else involves destroying and displacing what was there before, and not allowing other things to take place. Including, excluding, creating, and destroying the (natural and cultural) elements of reality raise moral issues that place helps initiate from the very beginning. The purpose of this book then is to explore what the geographic problematic means and to show that, when fully developed, it leads to insights about qualities of the real and the moral or good that are not as clearly visible through other approaches, and that can be used to evaluate and guide our desires and interests.

The claim that we can gain moral insight by confronting our geographical nature needs the support of a *geographic theory of morality*. This theory and its conclusions are offered as a provisional sketch of what I hope will be a continuing conversation about geography and morality. I am not claiming that this theory breaks new ground in moral philosophy. Rather I believe it sheds light on, and makes more accessible, several important relationships between morality and reality that would otherwise likely be overlooked. I am also not ignoring the fact that geography and morality have been connected before. Rather I am recognizing that until recently the connection has been filtered largely through religion. Since the dawn of time people tended to think that the earth was the gift of the gods and served a divine purpose. So very intertwined are these ideas with religion and mythology that the moral became embedded within them. Only in the last several decades, and most emphatically through the pioneering works of writers such as David Harvey, David Smith, and Yi-Fu Tuan have explicitly moral issues become part of geography's agenda.[3] This book is an attempt at a sustained and systematic argument about geography's role in morality. To be moral is to do the right thing, to do good, and this in turn means we have to have an idea of what is good, an idea of *the* good. Geography sheds important light on the real and the good without employing religious or theological assumptions. The good in this argument does not require the existence of God. God is not the issue; rather it is the good.[4]

The theory will show that geography helps illuminate what is universally, essentially, or, as I prefer to call it, *intrinsically,* good. The basic idea is that our place-making and earth-transforming activities should be guided by the *joint* application of two criteria that draw upon this intrinsic good: *we should create places that expand our awareness of reality*, and *we should create places that increase the variety and complexity of reality*. In doing so we would be creating better places and a better world, and would be moving in a moral direction. The joint application of these two criteria becomes our *geographical guide to the real and the good*. The application must be joint, for each counterbalances the excesses of the other; applying them jointly stresses the active and processive nature of doing good, and provides us with a direction that lies between moral relativism and absolutism.

If these criteria are convincing, as I hope they are, they then can become part of our image of what reality ought to be, guide us through the problematic, and help us to judge our transformations. These criteria do not specify what in particular detail such places should be or do. Rather they recognize that our efforts at transforming the world will be continuous and that there are innumerable paths that could be taken, but that they all ultimately lead in the same direction. Without such criteria for judging and guiding, our transformations can all too often lead us in tragically wrong directions. These two criteria are what the geographic theory of morality advocates as a guide, and it is through them that geography sheds light on morality.

What though is meant by the real and the good? Where does the geographic theory of morality stand regarding free will, moral absolutism, and moral relativism? These questions will be addressed in detail throughout the book, but in order to provide a sense of where the theory lies in relation to them, I will introduce briefly four of its most important sets of arguments.

First is the argument that to be a moral agent one must be able to choose. This means humans must possess some degree of autonomy or free will. Certainly, much of our behavior is conditioned or caused by forces that surround us. But for us to be moral agents there must be some, however small, degree of choice that is not itself caused by these conditions. That is, we ourselves are selecting (and not something in or around us causing us to select), for if our choices are caused by preexisting conditions and forces, then we are not responsible for our actions.[5] This point about autonomy and free will as the basis of morality would be countered by many who see themselves as moral naturalists.[6] To use an overly simplified example, a sociobiologist may argue that we are products of selfish genes. My care, concern, and self-sacrifice for my children, nieces, and nephews are felt by me to be good moral deeds—things I strive to do. But to the sociobiologist, they are acts that my genes force upon me in order for these genes to be perpetuated. My choices, and my belief that they are good, are determined for me—they are motivated by self- (or in this case genetic) interest. My point is that if they are really things I am forced to do (even though I may think I do them voluntarily), they are not necessarily moral, for I had no choice or responsibility in doing them. I do not deserve praise. Nor do I deserve blame for not doing them, since that would presumably be a result of a genetic defect. This difficulty about causality pushing aside the moral and making it simply an instrument for justifying or rationalizing actions that we ourselves do not select (though we may believe we do) is equally true when the causes or forces are thought to arise from other aspects of nature, or from underlying mental structures and meanings, or from our social and economic conditions. Again, I am not denying that these forces exist and are powerful, but that morality enters the picture at precisely the point where we have some degree of autonomy and thus can do otherwise.

Second are arguments concerning reality. The theory assumes a type of realism. It claims that reality exists, though we can never know it as it really is, and our models of it come from a variety of perspectives, including the arts, the sciences, and the views of different cultures, and from numerous engagements with the world. We also are contributing to and changing parts of reality through the places and projects we create. Indeed, the geographic problematic stresses that on the earth's surface the reality we experience is thoroughly influenced by our efforts at place-making. These human creations are as real as the ones we did not create. The territorial United States, or a classroom, is as factually a part of reality as is Mt. Everest, or the Atlantic Ocean. Reality includes the things we create, but reality is not simply a product of our imaginations, though we can approach it only through our models and concepts. Reality is infinite and infinitely complex, and our knowledge of it is always partial and provisional and comes from numerous points of view. Reality is also compelling. It arouses our curiosity. And of course, reality affects us. Still, its infinite and ineffable qualities make the possibility of comprehending it certainly and completely always beyond our grasp. It is like a continuously re-ceding horizon—we are always drawn to it, and know more about it, but we can never reach its end.[7]

The third set of arguments is more controversial, and so too are its implications. At the core, it claims that the good is part of reality and thus is also real. It is an independent part of reality that we do not make up, though our acts manifest it in infinite ways. This means we can be realists not only about empirical reality, but that we can be moral realists too. Recognizing that the good is a part or subset of real-ity that we do not invent or make up gives moral weight to the idea that human beings have free will. The existence of a good that is real provides individuals with something to choose that is independent of them—that is not simply something forced upon them by their ge-netic material or their class position, or something that simply ratio-nalizes their own interests or projects. The good, as a part of reality, does not cause us to do good. It is still our choice to be aware of it (and our lack of effort is the source of evil). This idea of the good leads to criteria for guiding our choices and judging their goodness that are not themselves determined entirely by the conditions that surround us. The "ought" in the problematic need not then simply be a product of the empirical conditions and forces that exist as is; rather the ought can be informed by an intimation of an independently existing "in-trinsic" good.

So, the good then is part of, but not produced by, the rest of reality. Also, not all of reality is good. And, as with reality, the good too is infi-nite in its manifestations and ineffable in our capacity to understand it fully. These limitations to our knowledge are expected since we are creatures who have evolved with specialized talents, making us adept at some things and not at others. As Colin McGinn puts it: "The nat-ural world can transcend our knowledge of it precisely because our knowledge is a natural fact about us, in relation to the world."[8] We thus have talents and blind spots and so our knowledge of reality will always be incomplete, and so too will our knowledge of the good. All we can ever have is an intimation. While there are infinite paths to-ward it and our actions provide diverse manifestations of what good can be, we do not make up or invent what makes these good. If we did not make these assumptions about the reality and independence of the good, but rather assumed it to be simply a product of factors that fit our own self-interests (a position that I will refer to later as an instru-mental one), there would also be little basis to our belief that we are to some degree autonomous—and to exactly this degree—moral agents. This is how the argument that free will is necessary for moral action is related to the claim that the good is an independent and ineffable part of the real, and that evil is due to a lack of awareness.

One important implication of the reality of the good is that it may allow the theory to help us navigate between two basic problems that plague moral viewpoints: moral absolutism and moral relativism. Both of these are undesirable positions. This may appear obvious in the case of moral absolutism, for it not only assumes that the good is real but that it is also unambiguously knowable, and that it can be summarized in a set of commandments. A danger of course is that these commandments may be wrong and yet adhered to absolutely. A more subtle danger of absolutism is that it narrows and trivializes the good. It makes it into something that is no longer liberating and expansive, but rather more like a means of oppression. By contrast, the theory's assumption that the good is real but ineffable—that it is a compelling but continuously receding horizon with infinite paths to it—avoids the dangers of absolutism that come with certain, fixed, and uncritical knowledge.

Still, one can argue that any moral theory can, even unintentionally, lead to absolutism because it can close off discourse, discourage different points of view, limit our receptivity to new positions, or simply be taken as dogma. I am extremely sympathetic to these concerns, for they derive from the generous moral impulse that encourages open-mindedness and tolerance. They suggest that there are many paths to the good, that no one can have the complete set of right answers, and that people must be involved in the process of deciding what is right and moral, and not have it imposed upon them. I share these views on tolerance and pluralism and argue that they can and should be part of moral theory—and that they are part of this one. But it is important to note that to abhor intolerance and absolutism is itself a moral position that supposes there to be a defensible conception of the good that is not entirely personal, and that may be shared by others. It is difficult to do this without articulating this idea of the good and its implications, and this means expressing them as clearly as possible in a moral theory.

To many, absolutism seems more objectionable than relativism, because the latter appears to be a route to tolerance and open-mindedness. But this is really not the case. Indeed, relativism can lead to absolutism. What though is moral relativism? In its simplest form it argues that the good or the moral is not real and independent but rather caused by natural or cultural forces: the moral is really a rationalization of mores, customs, and self-interests. There are as many "goods" as there are interests and projects that need justification, and so there is no way to say which one of these is better than another.

Your idea of the good is relative to and determined by your position, and mine is relative to and determined by mine, and so, as the contemporary expression goes: "You're o.k, and I'm o.k." This, though, does not lead to open-mindedness, but rather opens the door to intolerance and absolutism, because having someone claim that all morality is relative to the goals and practices of particular groups and interests means that there is nothing preventing one group from finding it in its interest to impose its morality on everyone else. To argue that such imposition is unjust or immoral is then not to be a moral relativist, but rather to have an intimation that there is a good that is not a product of any particular group—an intrinsic good that is real. This is why the idea of the good that is part of the moral theory developed here—that the good is real but infinite and ineffable—clarifies why relativism as well as absolutism should be avoided and helps provide a means of navigating between the two.

A fourth set of arguments is that the good is not only real, but that it is attractive and compelling. If we have an intimation of what is good, then we will be drawn to it. I have also used the words compelling to describe the rest of reality, but there is a difference between how the good and the rest of the real attract us. For the real, even its horrors and evils are things about which we must know and may even be curious. Yet, these negative qualities of the real are not things we want to emulate or have guide us. But this is exactly so with the good, for it is not only something that we want to know, but also want to be and to follow. The good does not force us, but rather attracts as would a lure guiding or beckoning us. This is what I mean by the good is compelling. A critical consequence of this, as we have already noted, is that evil arises from an insufficient awareness and clarity about what we are doing (and our responsibility and choice is to become more aware). If this is not true and if even when aware we find evil as attractive as good and do it willingly, then we are again led to moral relativism, but this time because there would be no difference between good and evil, the distinction would be completely arbitrary.

The claims that the good is compelling and that evil arises from a lack of awareness makes it clear that the site of free will and moral responsibility does not lie in accepting or rejecting the good for its compelling quality means we would accept it. Rather the role of free will and moral responsibility enters at the point of deciding to make an effort to increase our awareness. This argument that the good is compelling and that evil is due to a lack of awareness has a long philosophical pedigree that in one form another includes Plato, Augustine, Kant, Simon Weil, Iris Murdoch, and Hannah Arendt and her arguments about the banality of evil. Among the reasons the argument rings true is that it provides the only means of avoiding moral relativism while still allowing for free will and choice, and it is a position that squarely faces and condemns evil, and yet is politically progressive. It is also a position that makes the obligation to be aware partly up to the individual who must then choose to expand his or her view, become less situated, and see more clearly and thus not be subject to self-deception, and partly up to a society that must create and promote these opportunities. Through these efforts, the position allows for the possibility of moral improvement and progress.[9]

The arguments that the real and the good exist, that they are both ineffable, and that the good is compelling, point out how abstract and remote they are. Geography cannot make them any less so, but its two criteria for creating places—that places should help us expand our awareness of reality and should help create a more complex and varied reality—can provide guideposts or markers that are more accessible and down to earth. Like a compass bearing, these two can point us in the direction of the real and the good and can be used to evaluate whether the places we are creating and the projects they support are helping us move in the right direction. Without a good that is real, and without these guideposts, the geographical problematic would still operate, but without any direction. We would have no way of evaluating the places we create and the world we build. We simply would be creating and changing for no real purpose. And without these guideposts or compass bearings, moral theory remains too abstract, remote, and thin to fit real, concrete conditions and actions.

The belief that geography could in fact provide such guideposts to the good was expressed more than a century ago by the great geographer and social philosopher, Prince Peter Kropotkin, who, in his essay "What Geography Ought to Be," spoke about geography's potential for imparting a moral view: The discipline, according to Kropotkin, "must teach us, from our earliest childhood, that we are all brethren, whatever our nationality. In our time of wars, of national self-conceit, of national jealousies and hatreds ably nourished by people who pursue their own egotistic, personal or class interest, geography must be . . . a means of dissipating these prejudices and of creating other feelings more worthy of humanity . . . It is the task of geography to bring this truth, in its full light, into the midst of the lies accumulated by ignorance, presumption, and egotism."[10] A geographic theory of morality can be developed more fully now than in Kropotkin's time because we have a greater geographic awareness, in the sense of understanding the implications of our geographic agency and of how space and place operate.

These are the essential points that will be made in this book. The remainder of this introduction will set out in more detail the problematic; the relationship among place, the real, and the good; and the content of the geographical theory.

THE GEOGRAPHICAL PROBLEMATIC

I have observed earlier that it is our geographical condition to not accept reality as it is, and to continuously construct and alter places to transform reality, and to transform this transformed reality . . . and so on . . . in order to have it conform to an image of what we think reality ought to be. Here I would like to point out how this problematic implies qualities about humans that set them apart from the rest of nature. First I want to focus on the distinction it draws between place and space.

Place and Space in a Geographic State of Nature

To illustrate the relationship between place and space, consider a thought experiment that leads to what can be called a *geographical state of nature*, which can be taken as something like a geographical equivalent to the thought experiments of Hobbes, Locke, Rousseau, and others, who are seeking to isolate the effects of the social, or political, or economic. Suppose we are the first humans in North America, arriving with a culture and language. We cross the land bridge, move into the interior, and soon realize that there are no other people. Not only does the land seem uninhabited, it seems unending. Now let us use the term space here to label this vastness, this apparently unbounded expanse of land. Even though there are mountains, valleys, woods, and clearings, and thus variations in the landscape that we can call place, and even though where we are standing, our location, can also be called a place, there are not yet places in the sense that I am developing here in this book. Rather, if we simply stand, or move through space we are geographically naked and will remain so until and unless we delineate and bound an area that we can set aside and that enables us to control at least some of what takes place. This can be something as simple as a lean-to shelter, or a pit that is dug around the encampment to protect us from animals and things that move in the dark. Whatever name we give it, and whatever scale it occurs at, we are carving out a place from space by bounding and attempting to affect, influence, and control aspects of what takes place. And this place, with rules of what may be in or out, this place-as-territory, is what I regard as a place, and which I will argue is constitutive of virtually any and all human undertakings or projects. Without it, we are geographically naked and cannot survive. We would be humans without language or politics or economics.

So we begin to carve out a few small places in this sea of space—this natural wilderness. Now, leaving this hypothetical past behind, and jumping to the 21st century, we find that we have turned all of the land and also much of the sea and air—through jurisdictional rights, treaties about fishing, national and international airspace, and air quality management, and the like—into layers of continuously changing places. Historically then, from small islands of place in an unbounded sea of space, we now have created on the earth a virtually continuous and overlapping places in which there is practically no space left unbounded; yet we sometimes try desperately to preserve or create places that give the illusion of being space—wilderness areas which when one enters provide the sensation of not being bounded.

I hasten to add of course that our places cover only a very small fraction of the universe as a whole, and the rest remains a vast sea of space. But in this corner of the universe, this earth's surface, when we now create a new place, we are in fact replacing places, for there is practically no space that is not already part of place.

This thought experiment is intended to provide a picture of how we are place-makers from the very beginning, and how the bounded and controlled area of space—place in the sense I am using it—surrounds us everywhere because it is an indispensable tool for undertaking projects. This is because place has an effect and produces a value-added that cannot be replaced by or reduced to other things. This effect comes from its structure and dynamics (which will be discussed later in Chapter 2) and makes place an essential instrument in the transformation of existing reality into the new reality we think we want or ought to have. The places we create are conscious and planned efforts (though they most often do not turn out according to plan). Human place construction and our capacity to imagine are interconnected. But what is the motivation to transform? Why do we want to have a new reality, and then yet another new reality? Several issues about geographic nature must be addressed before I can propose an answer.

Humans as Part of and Apart from Nature

The first and obvious point I want to make is that humans are by no means the only instruments transforming the earth. The forces of nature themselves are far more powerful sources of transformation. For example, purely physical processes affect and alter the earth's orbit and tilt, the movement of continental plates, the eruption of volcanoes, and the amount of radiation that falls on the earth. Many other changes have been due to biological forces that are of nonhuman origins. These forms of life from the very beginning have profoundly transformed the earth. Bacteria and plant life changed the atmosphere from one dominated by methane to an oxygen-based system. The deposition of organic material has contributed to vast sedimentary rock formations and to oil and coalfields. Biologically derived calcium carbonate contributes to deposits of limestone, and to the endogenic cycle—the slow up and down cycle of material from the surface of the ocean to the underlying crustal bedrock and back to the surface; and living organisms in general are crucial parts of the cycles of carbon, oxygen, nitrogen, sulfur, phosphorus, the hydrological cycle, and to the formation of soil.[11] Living creatures adapt to the earth, and in so doing transform it, leading to the possibility of other life forms, and so the process continues. Nature, or the natural world, is constantly changing, and nature now includes not only the physical, but also the biological, and the dynamics between the two.

Given that nature itself, including other life forms, is transforming itself, what then sets the geographic problematic apart from these natural transformations? Some may argue that there is no difference, because we are part of nature and our transformations should be understood in that context. While we are indeed *part of* nature, we are also *apart from* it. The crucial distinction stems from the fact that humans are conscious agents who can imagine a direction that has not yet occurred and decide to pursue it. That is, humans can imagine, symbolize, and decide, for they have free will. This distinction is borne out by the fact that we do not morally praise or condemn changes that are products of nature. The impact of a large meteor hitting the earth around 65 million years ago may have caused the great extinctions of the late Cretaceous, including the end of the reign of the dinosaurs, but we do not think of the meteor as virtuous or wicked. Nor do we morally condemn the spread of the Pleistocene glacial ice sheets for creating mass extinctions in the high northern and southern latitudes. Acts of nature change the world, but they are blameless in a moral sense because blame assumes that the agent causing the event had a choice; it could have done otherwise. This is true even when nature causes immense human suffering. Certainly we may speak loosely and "blame" a tornado, a hurricane, an earthquake, for destroying lives, or we may blame the rot in our house on the infestation of carpenter ants, or we can blame the *Pasteurella pestis* and rat-flea communities for causing Black Death or the plague, but these cases we are using the word "blame" to say that something caused a misfortune, not that the agents of change had a choice and should be held morally responsible (unless we believe that natural acts are caused by malevolent deities or are divine acts, in which case then we should blame God for evil, which raises its own problems).

The issue is entirely different when humans are the agents. We too are part of nature. And like the rest of life we transform it. But we do so in a different way and by different means, which is enough to also make us apart from nature. Unlike a territorial animal that stakes out an area by instinct, we do so through forethought and reason (though our reasons may often be misguided). The beaver must dam up streams and build wooden lodges. As far as we know, it does not think about and discuss with other beavers the option of doing something else to the landscape. The beaver, and other territorial animals, can be said to be creating "place." But these again are different from the kinds we create. They are based on and directed by instinct, while human place-making has infinite and open-ended purposes that require abstract thought and moral choices. Place, in our sense of the term, depends on and reflects the use of our imaginations and capacities to use symbols. This makes our place and place-making so very different from that of other creatures that our use of the term place does not apply to them.

When we humans dam up a river we have in mind what areas will be flooded and we understand that flooding precludes other things from taking place. The dammed-up water forms a bounded area that we control—the dam being the only exit. But unlike the beaver and other animals, we can change the purpose of this or any other place. Our dam may serve to harness waterpower, but then be used to create irrigation systems, to have a store of drinking water, and to use the flooded area for recreation. We also can decide to destroy the dam and unbound the natural process; that is, we can turn our territories on or off. We are conscious of clearing a space and creating a place, and so place can be used to varying degrees to alter the landscape—to change what is mixed and woven together.

Unlike other creatures, we use place to not only transform our natural environment and make it habitable, but also to transform the culture we have created in order to make new culture. Because place already helps us weave together elements of nature and culture (even a wilderness area, whose boundaries attempt to exclude most human activities, is itself a cultural intervention in a natural process, thereby interthreading meaning and social relations with nature), our uses of place to make new culture also involves the further transformation of nature. In the 1960s we may have leveled a forest and appropriated farmland to create a new suburb; twenty years later, much of this is leveled to create a mega–shopping mall; and we can expect that this will hardly be the last change. When we close our eyes and picture what the future will be like, our visions may differ greatly, but most of us do not expect the cultural landscape to remain the same. If we want it to be just as it is, we must make an effort to keep it that way. And so, unlike other creatures, we are capable of imagining alternatives to the reality we have; in preferring imagined ones, we attempt to make them real. This capacity to imagine a world that does not (yet) exist requires a facility with symbols and forms of representation so that what exists can be held in mind and compared to images of what could exist. It also requires a will to enact what we imagine and represent. But this process can be absolutely futile, destroying our chances as a species, unless it is constrained by reason and an awareness of what is possible. Imagination must then be tempered by a sense of reality (the world as it is and as it realistically could be) or else these imaginings become purely and only fantasies; even worse, they turn into forms of delusional and self-destructive madness. We must then know how the world works, which includes how place works. Such knowledge is extremely difficult and always incomplete, not least of all because we are changing what is real. We dream up and make changes to reality, but we do not make up the fact that these changes exist as part of reality. Because we continuously change reality by realizing (some of) our imaginings, it is essential we think through whether these changes are good or bad.

But what of our original question: why though do we change reality? Why do we expend so much energy in transforming it? Why can't we accept it as it is? The answer I propose is that we change reality because we want to make it better. We believe we can improve upon reality. The question now is better in what sense? Is it better in that the new reality allows a specific individual or group to achieve its goals, and thus better in a sense that is instrumental and relative to these interests? This certainly describes what most would think is the case. We want to make reality better so that it conforms to our own particular and partial goals. But the point again—is this the only way it has to be? Can reality also be made better in a more general and intrinsic sense of better? I will argue that it both can and ought to be, and that this sense of better depends on the connection of geography to the real and the good.

GEOGRAPHY, THE REAL, AND THE GOOD

Before I consider this complex relationship, I must stress that place is something we construct. It is an instrument, and it is through our agency that it has an effect. Yet we cannot be effective or even undertake projects without place, for its effects as an instrument are indispensable and irreducible. This role of place is analogous to language. It too is a human product, requiring us as speakers and listeners. It enables us to communicate and thus has an effect, even if that effect is ultimately dependent on us. Similarly with place. Even though we create place, we do so because it has an effect. These effects are mutable, and often unpredictable, but they are dependent on us. Since this is a geographic discussion, we will often stress this effect by saying that "place does something," that "it has an effect," or that "it is a cause," with the understanding that it is "we" who are the ultimate causes in this relationship. In terms of judging places, we then can think of places as good or bad and call a place like a concentration camp or a slave plantation *bad*, with the understanding that we humans are making them so, but in turn, we could not have done these evil (or good) acts without particular uses of place. (Even in this moral sense there is an analogue to language. We often say that particular words can incite us to do hateful things, or that other words inspire us to do good. Of course it is humans who are the creators and users of these words as it is the case that it is humans who are the creators and users of place; but we could not provoke or inspire without them). Bearing in mind then that places are good, bad, or far more often complex mixtures of the two insofar as they are being used for good or bad purposes—and that these purposes could not be undertaken without these particular uses of place—let us consider the connections among place, the real, and the good.

We have argued that there is a reality, that the good is real, that humans have imaginations and choices, and that we transform reality by constructing places. At this level of sheer geographical place-making we are affecting the real, and we do so whether or not we are aware of it. Our place creation can not only change reality, but also expand or contract it in different ways. That is, our place-making can help enrich or impoverish reality.

Yet there is another level at which place affects reality: a place can make us more or less conscious or aware of reality. The two levels are interrelated. On the first level—of simply changing reality—we might say that thirty thousand years ago, the cultural landscapes worldwide, with their hearths, huts, and hunting-gathering territories, were few and far simpler than is the case now when the earth's surface still includes some of these, along with the innumerable places such as rooms, buildings, offices, farms, municipalities, wilderness areas, and states in which activities and projects take place. It may seem then that in sheer numbers and types of cultural places we may have increased the quantity of the real; but weighted against this are the cultures that have disappeared and the recent dangers we pose to biological and ecological complexity. To begin to answer the question of whether on balance we have increased the scale and quantity of the real, and enriched or impoverished the world, we must be geographically aware of how we have affected reality, and here enters the second level, for the places we have constructed not only affect what is real, but influence our awareness of the real, some increasing our awareness while others diminishing it.

For example, a few places are specifically designed to attempt to make us more aware of the world. Schools should generally try to do this, and universities even more so. Numerous other places, though not dedicated to increasing our awareness, can nevertheless do so, or at least not impede our efforts to gain a broader and more realistic view. Even so, there are places that provide an escape from reality. Places of fantasy, such as amusement parks and Disney Worlds, although as real and substantial as any other place, attempt to distract us from the conditions outside and inside, behind the scenes. These places of escapism are not in themselves bad, and can even be refreshing, but only if they do not mold our awareness. At that point we run the risk of substituting fantasy and illusion for reality. A more severe challenge to reality comes from places that constrain our view with opaque boundaries. Extreme cases of narrowing our vision result from places that censor information about the rest of the world and about what takes place within their borders, as did Nazi Germany and Stalin's Soviet Union, and from places that are secret, such as some military installations and bases in our society, and of course the death camps of Hitler's Germany and the gulags of Stalin's Russia. We may not even know that such places exist, and even if we do, we do not know what projects they contain and so are thereby denied access to that part of reality.

Place and reality form a complex relationship. Even though we make them, places are real, and increasing their number and types in one sense may increase reality. Yet, some places are constructed so that they contribute to or expand our awareness of reality while others may diminish it. Clearly an expansion of the real then is not only dependent on the number and variety of places, but on what they do to our awareness of reality.

All places are real and many of them complex, but are all of them good? Even though places possess many facets, one or a few can be so very effective in helping us create good or bad effects that they color the entire place. (The same is true for people. Even the worst of us may have a few positive qualities, but these are not enough to save us from moral condemnation.) From what has been said so far, it is clear that some places are better moral instruments than others, that all places are not as good as they can or should be, and that some kinds of places are used primarily for evil. Moreover, the character of a place depends on its relationship to other places in a system: concentration camps and slave plantations are not good, nor were the geographical systems that supported them.

We can of course make such claims about these places without a geographic approach. A geographic theory would be valuable if it could add to our understanding of why these are good or bad ones. This is exactly what I hope is provided by this work: a geographic theory that itself emphasizes qualities of the good and their relation to the real that may otherwise be overlooked and that then can be used to judge what is taking place. The theory's encouragement to create places that make us more aware of reality (that help us "see through to the real") and that increase the variety and complexity of that reality point out what evil places lack. More important though, the theory provides an ideal that can be taken up as a reason to improve places and to create even better ones. I much prefer to think of the theory in terms of inspiring an ideal, but the idea of judging, even though it may sound harsh, is an inescapable part of holding to an ideal. And judging does not have to be harsh. It can simply be a reminder that we are going in the right or the wrong direction, or that we can do better. The theory then attempts to answer the question raised by our geographical condition. It says that there are qualities of the good that a geographical awareness brings to light and that can be used to inspire and help mold our geographical imaginations and evaluate our geographical actions. What then are the main ideas of the theory?

A GEOGRAPHICAL THEORY OF MORALITY

I will discuss the theory in two sections. The first focuses on the two technical parts of judging and their relationship to facets of the good illuminated by geography. The second focuses primarily on expanding the implications of the ideals of the theory. It emphasizes the shift in moral awareness that accompanies the acceptance of the theory, and explains how this shift corresponds to altruism and altruistic gift-giving, and implies an acceptance of the possibility of moral progress.

Criteria for Guiding and Judging

The theory argues that geographic judgments come in two forms. On the one hand are practical or *instrumental geographic judgments*. On the other are pure or *intrinsic geographic judgments.* Instrumental judgments are a truncated and often misguided form of judgment. Intrinsic judgments are able to complete the judging process and tell whether the instrumental judgments are in fact morally supportable ones.

Instrumental geographic judgments do not require a correct intimation of the good. Consider a slave society that finds plantations to be effective instruments. Consider also that the slaveholders believe that enslaving others is a moral act because slaves are inferior and will benefit from the paternalistic control of the slavemasters. The plantation then is not only an effective instrument in the goals of slavery, it is also justified by slaveholders as a tool or instrument in doing good. That is, the plantation is seen as a good place. How do we evaluate this claim?

In light of another level of judging—an intrinsic level—the plantation owners' judgments of what is good are suspicious because they appear truncated and self-serving. Framed more generally in terms of the problematic, the circularity runs like this: we transform the world by creating places to undertake projects, and our judgments of these places come to depend on whether they are effective in helping us undertake these projects. So what is held to be a good or morally justifiable use of place depends on whether the place is effective for the project we like and support. If that is as far as it goes, then it is likely to be an instrumental judgment.

In this case the project is slavery, but instrumental judgments apply to other major projects and also to everyday uses of place, including our assessments of the effectiveness of the geography of offices or factories for the attainment of the goals of a business; the effectiveness of classrooms and schools for teaching children; and the practicality of a new suburban development in building a community. In each case we often make complex assessments of whether the geographic organization of rooms, buildings, and larger-scale places have contributed to the attainment of the goals of these projects. If they have, and if they are projects we value, then this use of geography is instrumentally good in the sense that it is effective. The evaluation takes into account the multiple and complex facets of place and space, but if the justification is ultimately driven only by whether the place satisfies our self- or group interest, then it is likely to remain only instrumental.

It is important to stress that the group that holds these goals may believe that their moral evaluation is not simply a rationalization, but rather is about the truly good. That is, the group may think its judgments are intrinsic ones. A slaveholder may truly believe that slavery is good. The discussion of intrinsic judgments will enable us to assess these claims further and help break this circularity and relativism so as to determine if an instrumental judgment is in fact intrinsically good. But let us return for the moment to instrumental judgments in everyday life.

The particular goals of homes, offices, factories, and schools do not usually raise strong moral questions, although important moral issues take place within them. Still, the moral arises directly from the fact that each place prevents other things from taking place, and that each place draws together elements of truth, justice, and the natural (why these three will be explained in Chapter 2; here let it suffice to say that place helps weave together these three virtues in particular.) When the only overriding concern is the goals of the project, then the meaning of these virtues can become twisted by these goals. We may not notice this or be troubled by it when these places are part of our own culture and we are convinced that our culture's values are good. Then these particular mixes of virtues would be judged in terms of whether they in the long run contribute to these goals. But what happens when we observe a culture whose goals are very different from our own? What happens when the goals are those of the slaveholders? Then we may think that truth, justice, and the natural are being twisted to suit their interests. They are being used circularly or, more generally, instrumentally.

The same point can be made in the case of Nazi Germany. Here virtually every place was geared to Nazi ends, not the least of which was the attainment of racial purity—which to the Nazis was thought to be truly good. Each place then was to be dominated by the goals of the so-called highest race—the Aryans. They were not to interbreed with the "inferior" races: the most inferior, in their view, the Jews, were to be quarantined in ghettos and exterminated in concentration camps. Each place then helped constitute this idea of what was racially natural and pure or unnatural and impure. The degrading conditions of ghetto life and the inhuman conditions of the concentration camps simply reinforced the image that the people in these places were inferior. The geography of Nazi Germany also affected the meaning of justice and truth. One could see this in ghettos and concentration camps, and also in homes and schools. The obligations and duties within the German home, the lack of hospitality extended to "non-Aryans," and the subjects that were discussed around the dinner table; the admissions policies, forms of obedience, and subject matter taught in Nazi schools all differed from the way elements of justice and truth were conceived of in most homes and schools in other places and times.

To create this social organization with its particular set of values required a complex and remarkably effective web of places. The activities within the home, the school, the office, the factory, the barracks, and the special activities within ghettos and concentration camps required mastering the uses of space and place at multiple scales. Geography was used effectively, and the concentration camp system, which, according to Daniel Goldhagen, "was the main instrument for the Germans' fundamental reshaping of the social and human landscape of Europe" was perhaps the most effective place of all.[12] Places like Auschwitz, the concentration camp system in general, and the entire geography established by Nazi Germany was instrumentally good. It was an effective and efficient means of attaining Nazi goals. The Nazis thought that these places were even intrinsically good, for they were effective instruments in what they seem to have taken to be moral projects.

Instrumental geographic judgments alone, though intricate and involving knowledge about the structure and effects of place and spatial relations, are relative to the goals of the projects and to the interests these represent. Even if these goals are then offered as ultimate or intrinsic ones, there is no way that instrumental judgments can tell if that is true, for within this system of judging, morality becomes equivalent to mores or custom. Moral relativists, who see no real and compelling intrinsic good, claim that after all is said and done, instrumental judgments are the only kinds of moral judgments. Custom, mores, and self-interest are all we have. There are no grounds above the instrumental that allow us to say that the schools, homes, factories, and offices in Nazi Germany (and their particular views about truth, justice, and the natural) were not as morally good as those in other places such as England or the United States, or that concentration camps were kingdoms of evil. Thus, when instrumental geographic judgments are taken as the ideal for geographic action and the criterion for judging, the question of the last part of the geographic problematic becomes circularly referred back to the conditions operating in the second part—that is, it says that the transformations justify themselves.

Before I argue that there is a way out of this relativism through another and more general form of judging (i.e., intrinsic judgments), I should say that the problematic itself does not require that geographers find this more general form. Rather, geographers can be content with instrumental judgments as the only possible kind, or remain neutral on this subject, and content themselves with simply disclosing and analyzing the implicit and explicit moral arguments offered by individuals and groups involved in place creation and contestation. That is, the academic geographer can simply show how moral claims are constructed and used, and not judge such claims.

My view is that if the analysis ends there, it tends to support instrumentalism by default. This is so because by not judging whether a particular justification of place is itself moral, the analysis will reveal only how the justification fits the context and thus make the justification seem to be no more than a product of self-interest. Also, ending it there will unnecessarily truncate geographical analysis, for it does not recognize the capacity of geography to offer a guide to help us out of the circularity of the instrumental and lead us to a fuller sense of the good. And, many geographers have not ended it there. They take sides when they discuss problems of injustice, the need for critical perspectives, the importance of emancipatory social movements, and when they condemn environmental degradation. In doing so they are condemning the bad without really articulating the good—an articulation that would only add strength to their critiques.

This is where intrinsic geographic judgments play their role. These extend and judge instrumental ones, much like Kant's Categorical Imperative establishes a criterion for judging moral maxims and reasons for undertaking actions. Intrinsic judgments recognize that all places are instruments and attempt to discern whether a place is being used as an instrument for really doing good—intrinsic good; or whether it is, as we have discussed, only an instrument for some other type of goal that can be mistakenly or misleadingly called "good." Calling them intrinsic geographic judgments does not mean that the idea of the good is good only for geography—that it is instrumental to geography. Rather the "geographic" in pure or intrinsic geographic judgments means that this kind of geographical awareness illuminates facets of the good in general, and that these can be used as a basis for judging our actions and our use of places.

One of these facets of the good that undergirds intrinsic geographic judgments is awareness. Drawing on the argument that the real and the good are compelling and that we do not do evil willingly but out of a lack of awareness, it follows that it is better to be more than less aware, and that evil is (again) a consequence of not being aware enough. The other facet is based on the related idea of the goodness of a plenitudinous reality: a diverse and more complex reality is better than an impoverished one. A rich and varied reality becomes more compelling, and it also provides different points of view from which to see reality.

These two facets of the good are widely held human values and by no means restricted to geography, but geography does draw our attention to them and combines them in a particular way. It is in this sense that geography illuminates them and that they can form the basis for intrinsic geographic judgments. Each is in turn multifaceted, an issue that will be explored in detail in subsequent chapters, but for now we will discuss them as single entities. They are combined and employed jointly by the theory to evaluate the instrumental judgments and uses of geography.[13]

The first intrinsic geographic judgment, based on the goodness of being more rather than less aware of reality, takes the form of encouraging us to create and value places that heighten our awareness of the real and that share this awareness openly and publicly. Knowledge of reality should not be a secret, but rather open and accessible to all who want to know. That is, it should be made a gift, given freely to others (a point that will be elaborated in the next section). Knowledge of reality is a good in both a quantitative and a qualitative sense. We must know more and also must know more clearly, penetratingly, and deeply. Both aspects of awareness are intended to be captured in the intrinsic geographic judgment of "seeing through to the real."[14] The second intrinsic geographic judgment, based on the goodness of a reality that is complex and varied over one that is simple and monotonous, takes the form of encouraging us to create and value places that increase the complexity and variety of the real. This is the intrinsic judgment of "the value of a complex and diverse reality," or "variety and complexity" for short.

Together they constitute qualities of the good illuminated by geography. The two must be used jointly to make judgments. One is not sufficient and needs the other for counterbalance. Their joint application encourages us to create places that expand and enrich the world and our understanding of it, and their application helps us assess the particular instrumental geographic judgments and the moral qualities of place. Even though a place may be highly valued instrumentally, an intrinsic geographic judgment may show that it is not a move in the right direction for it diminishes our awareness of reality, or it diminishes the richness and complexity of the world. The two facets of intrinsic geographic judgment correct and reinforce one another. We need a complex and varied world to attract our attention, and yet this complexity and variety should not cut itself off from the rest of the world, preventing those inside from seeing out, or those outside from seeing in.

They must be used jointly, yet it may be when a place is seriously deficient in one that it draws our attention. The Cold War Iron Curtain of the Soviet Union—but also the "Bamboo Curtain" of China and the still-existing "wall" around North Korea—are large-scale examples of how places can make it exceedingly difficult and dangerous if not impossible to see in or out and thus be aware of the world. This diminished awareness was (and, in the case of North Korea, still is) so severe that it has given rise to serious moral difficulties that likely outweigh the benefits that came from the fact that these are places containing complex cultures and social practices that have also increased the variety of reality. They have, but most would be unaware of it, and if we were allowed to be aware, the places would most likely be transformed.

Places of poverty, opium dens, and areas that cultivate superstitious life can all be thought of as contributing to the diversity and complexity of the world, and to those on the outside they may appear intriguing and exotic, but again, this diversity would be outweighed by the fact that such places are clearly curtailing awareness of those within; for them these places could feel like a prison.

The joint application of these values means that places must be transparent enough for those inside to see out and those outside to see in (where again "seeing" and "transparency" are shorthand for being aware of what is taking place). But places can be too transparent. The boundaries of place may be so porous that virtually nothing of any consequence is taking place. No serious projects can be undertaken. Open to constant inspection and devoid of real content, such a place does not contribute to (and more likely diminishes) the variety and complexity of the world.

Most places are morally mixed, for they contribute to neither aspect of intrinsic judgments very much. Some places are clearly better, in that they contribute to both. A university that lives up to its name should be a place that encourages its members to openly and publicly explore the world to its fullest, and to do so in a variety of ways. And some places are clearly evil for they violate one of these criteria to the point where the other cannot offset it. In this respect, the Iron Curtain was so excessively closed and secretive that this outweighs many contributions the place may have made to diversity. And a few places are evil because they violate both values. Nazi concentration camps and United States slave plantations at one scale and the Nazi nation and the antebellum South at another diminished for both victims and perpetrators the cultural variety and complexity, as well as diminished and, in the concentration camps, destroyed the existence of awareness itself.

The obverses of these intrinsic judgments then become ways of defining places of evil. The first obverse arises from the opposite of seeing through to the real, and occurs when a place is excessively restrictive, impermeable, and isolating or *autarkic*. The second one arises from the opposite of valuing a more varied and complex landscape and is expressed in a *tyranny* of one place over other types with the result of homogenizing place. The third stems from violating both intrinsic judgments and leads to constant transgression and *chaos*. Autarky, tyranny, and chaos are morally objectionable uses of place, and are the opposites of the qualities emphasized by intrinsic judgments.

These cursory examples of the logic of the theory and how it can be used to judge and guide do not do justice to the complexity of the examples. These issues will be taken up later; here it is important to mention that when we use intrinsic judgments to judge or evaluate a place, these evaluations are about its use as an instrument. But unlike instrumental judgments, which see the place only as an instrument for particular projects and interests, intrinsic judgments see the place as a possible instrument for doing what is intrinsically and generally good—good that is not dependent on an individual's or group's own self-interest, but rather good that is an end in itself. Although the theory judges, it does not provide a detailed list or set of commandments of what should or should not be done. Rather it offers a set of bearings to guide our actions and judgments along the right direction—toward the real and the good. It encourages us to see our activities as a process that ought to expand and enrich our world. It recognizes that places, like people, specialize in virtues. Particular places draw together different facets of truth, justice, and the natural. But the theory helps us understand when these differences are different facets of the complex qualities of the real and the good, or are misconceptions of these virtues. Most importantly, the theory provides a guide for improvement for all kinds of place. A particular home, for example, that nourishes the child's biological needs, but not its social and intellectual, is not necessarily a bad place, but it could be made better even through small changes that encourage the children to become more aware of the world and to care more about others. And since these adjustments can be made differently in different homes, the examples of moral improvement on the landscape become more varied and complex, and a heightened awareness of this richer and more varied landscape provides further incentives for improvement. In other words, variety and complexity, and seeing through to the real, when applied jointly, judge, guide, and animate our place-making. They encourage a dynamic place-making process that expands reality, and our awareness of it. Each criterion reinforces and stimulates the other.

Why, one might ask, does the theory focus on awareness—on seeing through to the real—when clearly a concentration camp can be condemned directly as a place that murders others, and a slave plantation is detestable because it diminishes human dignity and freedom? Isn't the infliction of human suffering the core of evil, and its absence good? This is a central question, and will be addressed at length later in the book; here I offer a brief answer with four parts. The first has already been mentioned: the theory holds that evil is ultimately a lack of awareness, and so if we wish to find the root cause, and the root effect of immorality, we need to consider awareness. Because the theory faces squarely the power of self-interest and how particular emotions such as greed, hate, and envy can not only fuel it, but truncate our vision so that we no longer see clearly, it turns to an expanded awareness as an antidote. Why do we kill others, treat them unjustly, humiliate them and take away their dignity and why do we degrade nature, unless we do not fully understand what we are doing? Yes, we are gripped by vices and by institutions that have their own defects, but the tightness of their grip is due to ignorance; awareness of these defects loosens their hold and enables us to avoid conditions that would again place us in their grasp. What other grounds are there that provide a justification for moral critique and hope for moral improvement?

The second is that if the primary moral goal is the elimination of pain and suffering, this, in theory, can be accomplished by immoral means. A dictatorship, for example, can provide conditions in which its members become pain-free, drugged, complacent subjects. Few would call this a moral society, for the individuals are not themselves making choices and aware of alternatives. Rather they are being treated as pets.

Third, a focus on awareness is central to moving us beyond only a moral concern with others which is expressed as the issue of justice and care, to a concern also with both the idea of truth and properties of nature. Justice is undeniably a central issue in morality and it has been the preoccupation of most contemporary moral theorists. But as we noted (and will elaborate later), it is only one of three virtues that geography brings to our attention as moral issues. The other two are the search for truth and our relations with nature. All three are part of what is real and good. Neither of the two is necessarily subservient to justice. As we shall see, it is through place that we can discern how truth, justice, and the natural are related, and how they can be guided by intrinsic judgments.

Fourth, the importance of awareness, and the problems that result when it is lacking, is one which geography readily illuminates. Indeed, it is central to the moral quality of place. Place has the capacity to either expand or contract awareness. Its boundaries provide both an inside and outside, and so can push our attention in either direction. But place also reveals how the movements are interdependent: there must be a balance between transparency and opacity. On the one hand, seeing through to the real means not being overly constrained by the boundaries of the place, or its practices. A relatively transparent or porous boundary can encourage this expanded awareness. On the other hand, projects that are worth undertaking, and that provide the world with complexity and variety, need to have a boundedness that allows our minds to be drawn inward to attend to these undertakings. We must focus on what we are doing, and this means that the boundaries of place be opaque enough to temporarily block out most of the world. Still, this inward-looking aspect of being in place is only a means to undertake projects that must in the end be open to public scrutiny and that contribute to both a more complex and varied reality and to an expanded view of reality. This is how the theory tries to balance transparency and opacity. It recognizes the need for both, and that both are linked to awareness in the sense of an inward and outward lookingness, but in the last instance, the theory emphasizes the overall importance of an expanded and outward-looking view. This emphasis is part of the logic of intrinsic judgments, but the same stress is even more discernable when we consider that to accept the theory and make instrumental judgments more like intrinsic ones is at the same time to become less involved in self-interest and motivated more by the outward impulse of altruism.

Moral Awareness, Altruistic Gift-Giving, and Moral Progress

Instrumental geographic judgments remind us that we are self-interested and situated creatures.[15] In one respect, intrinsic geographic judgments concur. Self-interest and situatedness will always be the case to some extent, but intrinsic judgments add that the degree of self-interest can be lessened as we come to recognize our capacity to be more altruistic. Altruism, in its purest form, is to act without concern for oneself, and, through these acts, to bestow altruistic gifts in the form of knowledge, care, and material objects, that are given away with no strings attached. That is, a purely altruistic gift is given without control over who receives it, how it is used, and without expectation of anything given in return. Unlike a gift given away because one expects to gain something in return, which is a form of "reciprocal altruism," pure altruism is simply giving.[16]

But here I must sound a cautionary note. For humans, pure altruism is almost a contradiction, for we are purposeful beings, and a pure gift given without strings attached implies that it was given without forethought. Yet to give is to intend something. And if there is no intention or purpose or forethought, then what is given has a greater chance of being counterproductive and even destructive. A person can invent a procedure that can use simple household products to make an extremely toxic and contagious disease and give this procedure away as a "gift" to everyone, with no strings attached. In a perfect world, we may cope with such an unwanted and destructive "gift," but since we are purposeful beings in an imperfect world, we must to some extent think about what we are giving or contributing. I say "to some extent" because much of the goodness in giving arises from the fact that we do not stipulate how it will or should be used, and the recipients are free to use or not use what is given and to feel inspired to contribute in turn. So even though we want a great degree of indeterminacy in the gift process, we cannot be motivated by good intentions and still give any old thing. For this reason (and others that will be discussed later) it is important to use the term "pure altruism" carefully and to remember that it is an ideal or goal, but also must in some sense be curbed by our best human qualities for it to be a truly worthy aspiration.

Given these qualifications, we find that pure altruism is a highly regarded ideal that motivates us to be less self-interested. It is embraced in aphorisms such as "it is better to give than receive," and undergirds many utopian models of life, including the attempt to condense socialist and Marxist principles of cooperation and sharing in the formula: "from each according to his ability, to each according to his need."[17] In a more specific sense, altruism inspires a care for distant strangers (others with whom one has no personal connection and hence nothing personally to gain) and it is at the root of a desire to share knowledge and beauty. Many would argue that examples of acts often cited as approaching pure altruism—as when Socrates died because he gave Athens the gift of intellectual integrity by calling accepted truths into doubt, or when Galileo endangered his position by advocating true Copernican theory that the Earth revolved around the sun; or when perfect strangers endanger themselves and their families by hiding and protecting victims of a holocaust—can be explained away as some complex and subtle form of self-interest. Such attempts to explain away the possibility of altruism are made because most biological and social science models of human nature assume we are driven *only* by self-interest (and its derivative—reciprocal altruism). They cannot accommodate a conception of ourselves as also guided by, and aspiring to, altruism. In contrast, the geographic theory of morality argues that both self-interest and altruism are real impulses and must be taken into account, and that our truly human capacities make sense only if we recognize the importance of altruism, and attempt to have that inform, and thus change in the sense of weaken, our self-interest—to have instrumental judgments become more like intrinsic ones.

The theory's commitment to the possibility and desirability of having altruistic goals (again, with the recognition that we can never attain pure altruism) is found in its assumptions, in the operations of intrinsic judgments, and in its implications for everyday practices. The theory assumes, for example, that we are not entirely instruments of forces; that we have free will and the capacity to decide or choose, even if almost all of our choices are constrained and dictated by circumstances; that this capacity requires, and makes sense of, other human faculties such as our facility with symbols and languages, for these allow us to hold in our minds at the same time the conditions that exist and the conditions that we think ought to exist, and thus make real choices possible. Altruism and the capacity to choose shift the geographic problematic from simply a description of purposeless and directionless change, to one that holds out the possibility and the obligation of change for the better—of a change that can actually improve what is real.

The theory's criteria of intrinsic geographic judgments themselves are linked to altruism. The obligation to have places help increase our awareness of reality—to help us see through to the real—cannot work if this awareness is in principle only accessible to some. Rather it requires that knowledge be shared—that it be made public and accessible to all who want it, and that there be few if any secrets. We cannot allow only some to know, for that would preclude the chance of increasing knowledge. Knowledge must be open to all and given away. It must be as open to inspection as is reality itself. To give away knowledge—to impart and share the truth—is often the motivation of an inspired teacher, of a creative artist, and of a research scientist, just as giving care and attention to others who are in need regardless of whether they can reciprocate is a motivation of those who are truly just. Yes, we must recognize that there is often instrumental or a self-interested advantage in sharing knowledge and helping others, but it would be a tragic mistake to ignore the reality and importance of an altruistic motive. We can be and often are inspired to impart insights about reality as a gift. The intrinsic geographic judgment of "seeing through to the real" assumes this is the case and that knowledge of reality must be as open and accessible as is reality itself.

This emphasis on altruism is sustained and reinforced by the second and equally important criterion of intrinsic judgment: that places should help us create a varied and complex reality. Variety and complexity not only draw our attention, but also are essential to creating differing views from which to examine reality. They provides checks and balances on what we see, allow us to sift, winnow, and discover new facets of reality, but again, only insofar as these different views are open to all who want to and can know. That is, they must be made public and given freely, just as reality is a *given*.

At the level of everyday practice, the theory encourages the altruistic impulse by reminding us that places (along with their products) ought to possess a giftlike quality. This attitude can affect our psychological condition, and our political and economic relations. For example, on the psychological level, following intrinsic judgments can help diminish the role of secrecy and self-deception. On the political level, the role of intrinsic judgments allows democracy to become more than a way of guaranteeing one's rights. It becomes a necessary process in sharing ideas and expanding awareness. And in the economic sphere, creating places guided by intrinsic judgments shifts attention from what might be called exchange-values, to altruistic gift-values. Unlike exchange-values, altruistic gift-values do not seem to be subject to the law of diminishing marginal utility.

These and other implications will be addressed in the following chapters, after the arguments about instrumental and intrinsic judgments are made. Before I outline the topics to be discussed, five cautionary notes need to be raised.

Cautionary points   One is that a moral theory cannot force us to behave well. It can only persuade through logic and reason that it discloses a better way of acting. Still, we must always be skeptical about any moral argument and accept it only provisionally and critically (as should be the case for those readers who find the arguments of this book persuasive) lest these arguments lead to a form of absolutism. Still, suppose a moral theory is found to be reasonable. Then we not only provisionally and critically entertain it, but also come to view the world by its tenets and take up its principles as a guide to our own actions. This again applies to the geographic theory. If it is convincing, it should be considered in constructing and maintaining places. As the instrumental uses of places are guided by the intrinsic, they become more like it. Transforming the instrumental into the intrinsic, or having the intrinsic absorbed by the instrumental, is how the theory conceives of moral geographic progress.

Progress will occur more readily if the theory is sound and used as a guide to action, but this is not to say that this or any other moral theory is necessary for good things to happen. Many people are and do good without being the slightest bit aware of moral theory, which also means they have constructed good places. In such cases, if the theory is of value, it would illuminate why the places and acts are good. In other words, it would show that the places allow us to see more clearly and add to complexity and diversity. So, moral theory can help clarify to those who have done good deeds what it is that they have done, and it can serve as a guide to the rest who need assistance in knowing what is good.

The second note is that this moral theory attempts to make explicit what has animated much of geographical inquiry. While the Kropotkin quotation I cited in the beginning explicitly claims a relationship between geography and morality, much of contemporary geography has shunned discussion of the moral. In part this is because some believe that developing ideas of the moral is synonymous with limiting the possibilities of tolerance. By now, however, it should be clear that this moral theory not only values tolerance (which itself is a moral position) but also tries to clarify many of the moral concerns that animate much of contemporary geography. Geographers who are interested in critical theory, feminist theory, postcolonial theory, poststructuralism and deconstruction, and of course Marxist theory, may be more comfortable discussing social justice than moral theory. But concepts of justice are themselves part of moral concepts and theories, and need to be justified by ideas that draw on the real and the good. More generally, to argue that something is unjust means that we not only have an idea of what is just, but of what is good. The present work is an attempt to make these assumptions and their consequences explicit.

The third point is that I wish to differentiate between the altruism in this theory, and the more familiar idea of reciprocal altruism. Altruism here involves a willingness to be less self-interested or selfless. I do not believe that this kind of altruism is possible to engineer by expanding our networks of interdependencies. Nor do I think the results of the two are the same. I am not denying that an increasing network of reciprocal connections among one another can expand our circle of care, concern, and gift-giving to include distant strangers. Certainly, social networks and webs of gifts that extend outward in space and time make it in our self-interest to give degrees of knowledge and care away to distant strangers, for they eventually can do things with it that will benefit the giver (which is the idea behind Mauss's *The Gift*).[18]

While I believe it would be a good thing if there is in fact a tendency to increase our care and concern for others as our web of interconnectedness increases (so that we come to realize that in a densely interconnected world "what goes around, comes around" and thus it becomes prudent to "do unto others as you would have them do unto you"), this idea of an expanding web of care resulting from enlightened self-interest is not based on a real or intrinsic moral concern; rather it is driven by empirical conditions for which we may not be responsible, and that can change at any moment, and thus change our range of care. It makes morality a result of what is, and not what ought to be, and thus another form of instrumental judgment. This is why I believe that enlightened self-interest or reciprocal altruism is not the same as altruism, and why it always needs to be reinforced by a willingness to be less self-interested and more altruistic.

The fourth point is that we and the places we create are imperfect. There may well be a limit to how much instrumental judgments can become like intrinsic ones—a limit, in other words, on how good places can be. The limit stems primarily from the dual facts that even the best place must bound our interest and focus our attention inward, as well as allow us to see out, and that even if we intend to be good, we will still devote much of our lives to escaping awareness and do so by creating degrees of opacity around our places. These are facts of life which a theory of the good must address for it to be informed by the real. While these facts may limit what we can do, we must still try to overcome them, for if we succumb to them we will likely live in a contracted moral world that sees right and wrong only instrumentally.

The fifth concerns the role of examples along the "continuum" of good and evil (which is in quotation marks because there again are innumerable paths to these ends). Even though evil is due to a lack of awareness and thus is an incompleteness, it seems to be a fact of life that examples of it are most always more sharply in focus than are examples of good. In a work of fiction, a person of bad character is usually more vividly and interestingly portrayed than one who has no flaws.[19] This is also the case in portraying places. Dante's description of Hell is more interesting than his depiction of heaven. The same is true in the visual arts. Scenes of cherubim and angels floating in a heavenly ether are rarely as exciting as the renderings of human horrors and sufferings. The human imagination can portray worse worlds more vividly than better ones, which is also true for this book. Its title is about the real and the good, yet the good side of the continuum has no end because the good is ultimately ineffable and so there can be no examples of perfect places (or people); rather the only thing that can be offered is a guide or compass bearing that points in this forever continuing direction and some examples along the way.

Not surprisingly, most of the examples in this book are about bad places such as concentration camps and slave plantations. They are not fictional renderings, but real examples of the most evil types of behavior that humans have enacted. Indeed they equal or even surpass what fiction and painting portray. In this respect they serve perhaps as an end point for the evil side of the continuum—places with a virtual absence of awareness, an absolute zero on the moral scale. Although they are vivid and attract our attention, when seen closely they become revolting and help move us in the other direction. Another justification for a focus on such places is that if a moral theory has any merit at all, it must shed light on why these extremes are so bad.

But what about the numerous actual and possible places that lie between this end point and the ineffable good, especially the ordinary places we take for granted? Here I will pepper the discussion with examples, but I will not do so in a sustained and systematic way, for these places are always shifting along the continuum. In an enlightened and generally well-functioning society, a part of that society, a home for example, is supposed to be a place that sustains and nourishes human life. This is good. But, as we all know, not all parts of even a good society function at the same level. A particular home even in a good society can also be a site of domestic violence and child abuse. If these were to stop, the home then would be a better place. And if the home were to create an environment which supports more than the biological needs of the individual by also encouraging open discussion and concern for others, then it can be even better. And different homes can do so differently; some may encourage a greater intellectual awareness and become more like schools, and others may encourage a greater social involvement and care for others. And in creating these different paths we are also creating a broader and richer understanding of what the home can become, and each of these can serve to inspire other and yet different attempts.

Or, consider another ordinary place like a restaurant, offering inexpensive, quick, convenient, and well-prepared food. On the face of it (and with due consideration to the superficiality of consumerism), this can be a good thing.[20] But suppose the place provides questionable working conditions for its employees. It becomes better if it attends more to the needs of these employees, and it can become better still if all of those involved in the organization have knowledge about the operations of the place and its connections to the larger world and are part of the decision-making process; and it can even become better yet if those involved in the place strive to make themselves and their customers aware of where, and the conditions under which, the produce being used have been grown and transported so that the act of preparing and selling the meal, and the act by the consumer of eating it, become also a geographically reflective one that extends our understanding of the consequences of our actions. The landscape we create can become better still if we increase the variety of ways in which this awareness is expended. Each type of restaurant can do so differently, and simply having a variety of culinary styles enriches and expands our experience of reality. One that serves Mexican dishes can, if nothing else, make us aware of the creative uses of raw materials such as corn, while another that specializes in Chinese food, can make us understand the role of soy as a basis for cooking. And, if this and other types of ordinary places were to open up their records and bookkeeping, encourage collective, open, and participatory labor practices, and encourage self-conscious links to other places and projects, then the variety and complexity and awareness of the world would be expanded still more, which could eventually transform the original systems of production and consumption.

Through such incremental changes in the use of place, even the most ordinary taken-for-granted places (and ones that even try originally to disguise what really is taking place) can become better, and it is through such small acts that we can cumulatively move along the continuum to the good. So, while there are no examples of the extremes of the good, there are numerous and modest ways that goodness and moral progress can be exemplified. We can even make such progress by taking places that were once used for evil and have them now expand our horizons. This occurs when the slave plantation and the concentration camp are turned into sites of memory that make us aware not only of the evil acts of others, but of the evil that we too can commit.[21]

A final point about examples. Since places are necessary instruments for undertaking projects, we cannot adequately understand a project without considering its geographic foundations. Slavery, fascism, anti-Semitism, totalitarianism are all projects, though they are more particularly political or ideological ones. To see how they work, we must examine how they use place at a host of geographical scales and levels, and especially the places in which these projects are most distilled—the gulags, the slave plantations, and the concentration camps. It is an advantage in presenting a moral theory to use such well-known projects and their sites because it saves time and also provides an efficient means of comparing the views of this theory with the more conventional nongeographic ones. The theory should also be used to analyze religious fundamentalism and intolerance, and it should be applied to projects that are "closer" to home. Poverty, race relations, gender inequality, environmental degradation, abortion, drug policies, criminal incarceration, to name but a few, ought to be examined in light of the theory. To do so requires a painstaking uncovering of the role of places and their interconnections in the formulation and execution of these policies, followed by the use of intrinsic geographic judgments to evaluate them. I trust the projects examined in this book will serve as a template for how these other policies can be examined, and will also provide good reasons for why such analysis should be undertaken.

The problematic presents us with the classic is/ought distinction: is the moral determined by the empirical conditions that surround us, or is it based on something else—a good that is real, compelling, independent, and ineffable? In either case, geography plays a central role. If it is the former, then morality is purely instrumental: it is mores or custom. Part or all of the empirical conditions that affect our behavior—our biological nature, our political and economic environments—then produce the moral, which becomes a rationalization for these conditions. If this is the case, the structure and dynamics of place, in the hands of instrumental judgments, explain how it is a means of weaving these empirical conditions together without reducing one to another, and how the thickness and density of these weaves hold us in their grip and relativize our moral views.

The second choice, and the one argued for here—that the good is real, compelling, ineffable, and independent—presents the case that the good is not a product of the empirical conditions surrounding us and that are woven through place. These empirical conditions certainly influence how easy or hard it is to do the right thing, and they make it possible to do good in many different ways, but they do not determine what it is. The good is independent, and we must rely primarily on the less tangible things such as our free will and our capacities to reason in order to have an intimation of it. This intimation then will animate how we create places and organize our material world so that these in turn can assist us in becoming better. I argue that intrinsic judgments help us to perform this function. Having in mind that our place-making should jointly increase our awareness of reality and increase the variety and complexity of reality can serve as a guide to a better world. Intrinsic judgment then can reshape our instrumental uses of places along these lines so that the weave of truth, justice, and the natural are now informed by the good.

The compass bearing provided by intrinsic judgments helps direct us to these intangible qualities of the good. In that sense it can make them more accessible. But we can never reach the goal of being good. Any example of goodness can be improved upon. It is also hard work to try to be good, in spite of the fact that the good is compelling. And, though evil is a lack, it is still more vivid, tangible, and accessible than the good. There is nothing we can do to reverse this. The moral theory, through intrinsic judgments, can help offset it to some degree, but even in a book that is about the good and is optimistic about the possibilities of moral progress, we cannot escape the vividness of evil. Its shadow, through its examples, will be present even in a book that does not intend to give it equal billing.

THE LINE OF ARGUMENT

The points made in this introductory Chapter 1 are reframed in the next Chapter 2—The Power of Place—by showing how place is central to both instrumental and intrinsic judgments. There I explain exactly how place's capacity to weave together elements of nature, meaning, and social relations, without reducing one to the other, makes it an indispensable instrument for undertaking projects. The argument draws on and extends my previous work on the structure and dynamics of place, and can be taken here as a theory of place.

Having argued for the centrality and efficacy of place, the next two chapters examine the role of place in instrumental judgments. This forms Section I. Instrumental judgements claim that morality is a product of context and place—it is a form of situated knowledge. My point is that this view, by itself, cannot avoid leading to moral relativism, or moral absolutism, or to both. The first part of the section, Chapter 3, considers how the situated argument is intimately tied to theories of relativism. I select several theories, including postmodernism and part of Marxism. Postmodernism assumes things are irrevocably situated, though some strands of postmodernism wish it were not so. Marxism is far more complex, offering a view of humanity as free and autonomous, but yet overwhelms this view with a weighty empirical theory that argues for situatedness insofar as class is a determinate of our actions and moral views. According to this important strand of Marxism, morality becomes a product of, or an instrument for, class interests. Morality is situated and relative. Whatever the form, arguments for situatedness and relativism are logically unstable and lead to difficulties in practice. One cannot live consistently as a relativist, and so Chapter 3 examines only theory. The unstable quality of relativism allows it readily to slip into absolutism, and here we have many examples. Chapter 4—Situatedness and Absolutism—examines the instrumental use of place in three cases: Nazi Germany, Stalin's Soviet Union, and slavery in the antebellum South. (Because intrinsic judgments build on and redirect instrumental ones, much of the evidence presented in this section will be reexamined in the next section on intrinsic judgments.)

By far the most attention is given to the case of Nazi Germany because of its abundant documentation about the motives of its leadership. I only sketch the cases of the Soviet Union and slavery in the U.S. with the understanding that the use of place in the Nazi Germany case could stand as a model for these two and for others, given the data and the time.

The weight of argument in Section I then is that instrumental judgments do not offer us much hope of escaping absolutism or relativism. Section II presents intrinsic judgments as a means of completing the instrumental, rescuing it from relativism, and doing justice to the problematic and to our capacities as moral agents. Instead of the moral being a product of the empirical, it is now seen as animating it. Chapter 5 develops and elaborates the moral theory and the role of intrinsic judgments by examining the way in which the judgments allow us to balance the need for places that are varied and opaque enough so that valued practices and ways of life can occur, and yet that are open and transparent enough so that we can see through to the real and make our views open and accessible to others. It also explores how the real and the good are compelling but in different ways. If the good is compelling, then the locus of responsibility for evil must reside in self-deception, which is addressed in this chapter. If we are not self-deceived and are able to apply intrinsic judgments, then we are also able to become less self-centered and more altruistic. It is through this capacity of altruism that truth, justice, and the natural—the key virtues woven together by place—are saved from relativism.

No matter how much we stress the role of intentions, good and evil are also a matter of actions, and both intentions and actions are shaped by the most basic qualities of our place-making—its boundedness. Chapter 7—Geopsychological Dynamics—explores connections between the material boundedness of place, and how it can on the one hand narrow our vision and move us toward forms of psychological compartmentalization that encourage moral drift, self-deception, and evil, and, on the other, help direct us in the other direction to an expanded and caring view. I discuss these dynamics and then use them to explain the process behind the absolutism and evil discussed in Chapter 4. Here I make the case that the link between compartmentalization and self-deception explains how slaveholders and Nazis did what they did. They were involved in a complex web of massive self-deception. This is the case even for Hitler. The last section of the chapter discusses how facets of geographical compartmentalization can be used to move us in the opposite direction—to expand our awareness and point us back to the good.

The moral aspects of place-making are not simply linked to the psychological (which is part of the realm of meaning) but also to the social. In discussing the social in Chapter 7—Geosocial Dynamics—I stress that it cannot engineer the good—it does not produce it and make us good. Rather it can make it easier to be good and helps sustain it. But we must first have an intimation of it and want to do it. So to keep our emphasis on the primacy of the good, I assume we accept intrinsic judgements and then sketch the most general geosocial relationships that follow from it. In other words, this chapter does not tell us how, socially, we move to a better world; instead it tells us what the social relations would be if we had such a world. I do so for the social (in the sociological sense), where I focus primarily on community and geographical scale; for the political where I offer what I call intrinsic democracy and contrast it to other forms including radical democracy; and for the economic where I offer gift-value, and contrast it to use-value and exchange-value.

This then is the flow of the argument, and the Postscript—Chapter 8, The Problematic and Moral Theory—suggests other implications to moral theory that is geographically informed.

2

THE POWER OF PLACE

INSTRUMENTAL AND INTRINSIC JUDGMENTS BOTH build upon the fact that place is essential because of what it helps us do—because of its effects. This does not mean that place does things on its own. Indeed, as we have defined it, place does not even exist on its own, but requires us as agents. Yet, it is an indispensable tool because of its consequences or effects. Both instrumental and intrinsic judgments accept this, but differ in how they interpret it. This chapter focuses on those aspects of the workings of place—its structure and dynamics and its relationship to space—that both types of judgments use, though differently. In the last section I will show how they differ. Why place matters involves a theory of its own which this chapter summarizes and molds to the arguments of this book.[1]

The best way to model how place functions as a tool is to think of it along the lines of a loom. As something like a loom, place helps us weave together a wide range of components of reality. The weave itself is the landscape and the projects that the place helps support. What does it draw together? In other publications, I have argued that the major components (or spools of thread) come from three *domains*: the empirical, the moral, and the aesthetic.[2] All of these are part of reality, and place helps us weave these empirical, moral, and aesthetic domains together. All three will be addressed, but I will focus on the moral and empirical because the moral is the center of our concern, and it is most often discussed in relation to the empirical.

Each of these domains consists itself of three *realms*. The empirical domain includes the realms of meaning, nature, and social relations;

and the moral domain includes the realms of truth, justice, and the natural. At the same moment that place as a tool helps to weave together the elements of these realms, it is thereby helping to weave together the domains. But for the purpose of illustration in the following figure (Fig. 1), I have simply separated this unified process into two planes,

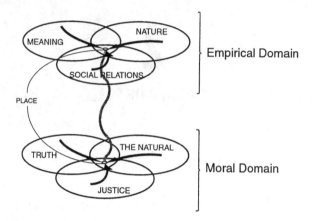

Fig. 1

indicating how place helps weave the realms of each domain. The two center dots represent the same place, indicated by the gray line connecting them. The curved arrows in each indicate that the place is drawing together threads or elements from the realms (the large circles) of meaning, nature, and social relations (for the empirical domain), and truth, justice, and the natural (for the moral). Without the agency of place, there would likely be no overlap. The weave that occurs within the place can be called the landscape. Even though the diagram splits the effects of place into these two planes, they are to be thought of as occurring simultaneously in the same place.

We must postpone the discussion of why the empirical domain contains the realms of meaning, nature, and social relations, and why the moral consists of truth, justice and the natural. For now, let us think of place as providing this loomlike tool that can help us draw all of these together and weave them into a pattern or fabric that constitutes the landscape and its projects.

The critical issue is the loom (which is not represented by this diagram—only the threads it weaves are). All places consist of this loom, though each type of place uses it to weave different strands and create different fabrics. How the loom works then is the core of this chapter. I will begin with descriptions of the threads and how different places help produce different weaves or fabrics, and leave to the middle of the chapter the core of the loom's mechanisms. In addition to the mechanism of the loom and the elements or threads it weaves together, there is a third issue—the individuals or groups doing the weaving. The questions "who determines the weave?" or "who is in charge of the loom?" or "who is pulling the strings?" are immensely important, though secondary to explaining why the loom is worth controlling, and will be addressed in the third part of the chapter concerning the geographic implications of the loom. Who is in charge of place matters only if we are convinced that being in charge matters. Unfortunately, much of the work in geography has focused more on who is pulling the strings and not on what pulling them can accomplish.

So, in part one of this chapter, I will discuss the threads and the weave by first addressing the way place assists in weaving together the moral elements of truth, justice, and the natural. Then I will focus on how place also help us weave together the empirical elements of meaning, social relations, and nature, and discuss the connections between the moral and the empirical. After that, I will discuss the heart of the argument—place as a loom, and then turn to who is doing the weaving. The weave and the loom constitute the structure and dynamics of place. Part two of this chapter will be a summary of implications. Some of these will be shared by instrumental and intrinsic judgments. Others draw attention to the limitations of instrumental judgments and suggest the need for intrinsic ones. Part three will focus on how intrinsic and instrumental judgments differ.

STRUCTURE AND DYNAMICS OF PLACE: ITS LOOMLIKE QUALITIES

Before I begin the discussion of place and its moral weave, it is important to realize that many places can be used to produce almost identical weaves. These are often simple purpose-built places designed mostly for one type of project—scientific laboratories, motel rooms, modern public schools, or, with some exaggeration, even complex places that have come to look more alike, such as the downtown business districts of modern Western cities. Designing places to be alike cannot completely eliminate variety, for we can never control everything that takes place, even in a small and intensely regulated environment such as a laboratory. Furthermore, the weave depends in part on the things at hand, which in turn depend on the relationship of this place to others. Still, we often try to create generic-looking and functioning places. By the same token, we also create places that are intended to be quite distinct. Distinctiveness will happen in any case with vast and complex places, such as nation-states, that contain multiple looms and weavers and complex unintended consequences.

A weave, whether it is similar or different from others, forms the appearance or landscape of a place, and also becomes a component in the particular projects it supports. The weave of nature and culture is clearly visible on the landscape of a farm, but it is also part of the farm's products: a cow, for example, bears the mix of these elements, and cows can survive and reproduce only in the kind of farmlike environment that molded them. The projects of a place—in this case a cow—bear this impress.

I am focusing on place, but of course we must remember that a place is part of a system; what one place is like depends on others. A soup kitchen is a place that arguably is concerned most with justice and care, especially the strand that refers to need. But it also is one among many types of places that do that. There are halfway houses, homeless shelters, emergency wards, and so on, and these are embedded within other types of places in a larger system. Moral specialization and interconnections among places allow for further specialization and also affects the moral meaning of these places. An unjust economic system may have led to unemployment, poverty, homelessness, and need. A soup kitchen then not only dispenses a form of justice but its existence stands as a silent internal critique of a larger social injustice.

It must be borne in mind that there is rarely complete agreement about what places do and what virtues they may or ought to possess. Places enhance human power and interest, and are constantly contested and in flux. The point is that the same loomlike structure must be employed by all sides, but used to different ends. Effectively undertaking a project—which can involve creating a new place or challenging what takes place, constructing or deconstructing it, making it more closed or open, more transparent or opaque, making it a site of oppression or liberation, or doing anything at all—requires activating the loomlike structure; we are powerless without it.

The Moral Domain and Its Realms of Truth, Justice, and the Natural

A house, a city, and a nation-state are places. But so too is a gulag, a concentration camp, a slave plantation, and an abusive home. Not only are these places, but unjust ones. Justice (or injustice) is a moral quality or virtue readily associated with place, because a place's rules and boundaries help or hinder our efforts to be just. But place is equally important to two other moral qualities or virtues—that of truth and of the natural. These are as important in our moral concerns as is justice. Seeing the world clearly, telling the truth, and not deceiving oneself and others is a central moral quality, for it opens up the possibility of awareness. And so too is the natural. Claims are made that it is important to be natural, to follow one's nature, and also to protect nature, for nature itself is a source of value. What these claims really mean is complex, but I will show that truth, justice, and the natural are related moral realms of virtues, with place as a central factor in drawing them together (Fig. 2).

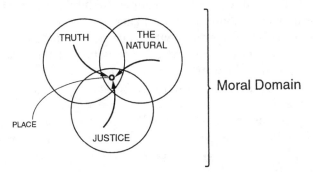

Fig. 2

Each of these realms of virtues names a huge category of often contradictory and contested qualities. Take justice. Three often discussed and difficult to reconcile strands are that justice can refer to treating people equally, or treating them according to merit, or according to need. Equality, merit, and need, are important elements of justice, but difficult to reconcile.[3] The same complexities and conflicts can attend the strands that compose the idea of truth. Truth can be based on how our ideas correspond to reality (that is, to a correspondence theory of truth), how internally consistent they are (a coherence theory of truth), or, on the degree of consensus that exists in a group, on what facts people believe to exist (a consensual theory of truth).[4] The natural too contains multiple meanings. It can refer to virtually everything that exists (to reality itself), or to the subset of reality that would operate without our conscious decision-making (which is roughly the meaning we will focus on in this book). Each type of place mixes—or weaves together—different strands and proportions from each of these virtues, but all places contain elements from the three. Here I want to illustrate the range, subtlety, and highly contested and changing quality of this weave. Many places specialize in emphasizing one or another of the virtues to the point where one or another of its strands becomes the controlling moral factor, with elements of the others playing supporting roles. On the other hand, there are places that tend to be less morally specialized, with elements from all three having center stage. This can be the case at all scales, but especially at the larger ones which contain numerous smaller places and many more weavers. Even then, large-scale places like a nation-state can attempt to create a particular fabric that stresses one or more virtues: a country may claim to stand for principles of liberty, equality and justice for all; or it may stand for the perpetuation of racial purity. True, there will be other components, but these may be the dominant ones purported to be in the weave. I will discuss later these larger-scale places and the relationships between what is claimed and what actually happens—the surface and the depth issue; for now I will continue with smaller-scale examples because they illustrate the structural components of place more directly and simply. They allow me a more fine-grained explanation of how place works. But what applies to them, applies to all scales.

Even though places are constantly contested and changing, there is, if not some degree of consensus, often an officially or publicly declared expression about what should (and even does) happen in a place. These conventional meanings may not mention any of these virtues explicitly, but they emerge as soon as we begin to ask what moral principles are supposed to lie behind the place. These conventional meanings are usually the points over which debates begin, and they will also serve as our point of departure to illustrate how place joins together facets of the three virtues—truth, justice, and the natural. My point is not to say that a particular place is in fact what it is claimed to be. Rather it is to point out that the claim and the counterclaims area always framed in terms of the realms and elements. I will begin with examples that focus primarily on truth and that use justice and the natural in supporting roles, then move on to examples of justice, and then to the natural. (The thicker arrow indicates either greater emphasis or more strands from that realm (Fig. 3).)

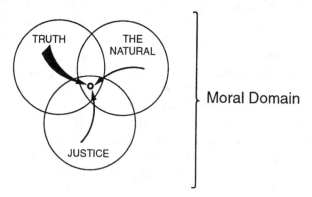

Fig. 3

Truth

Consider a lecture hall within a university. Both are part of other systems of places. Suppose it is a lecture hall within a university in a

Western democracy, as opposed to one in Nazi Germany (or in Stalinist Soviet Union) where there may be quite different mixes and meanings of the virtues. Consider also that the lecture is about geography, and specifically about characteristics of place. In such a place, within a society whose ideals are free and open intellectual inquiry (and whose university mottos include pledges to *veritas*, to *numen lumen*, and to "sift and winnow so that the truth can be found"), one would expect that the primary moral issue guiding the discussion is the virtue of truth. The professor is not supposed to be telling students just any old tale. She is doing her best to discern qualities and characteristics that places possess. She does not make these up; rather she expects that her lecture is disclosing things that are really there, or at least uncovering approximations of what are really there. This is not only the case for her own ideas about place—her own model—but for her review and discussion of what other people have said. Even though she may claim that their models are flawed, she must portray them as accurately and fairly as she can. She may also wish to explain where they are mistaken. But again, her explanation is offered as a statement about what she thinks are their failures. She would not (deliberately) distort things or make them up. She is motivated by a quest for the truth.

If she were to think about and articulate her conception of truth, it might be along the lines of a correspondence theory. This means that she believes there is a reality, and places are part of it. Places possess a structure, and her job is to come as close as possible to uncovering and representing it. What she thinks the structure is like could be thought of as *her* model of that part of reality. This idea of truth and the quest for it, is also likely to be what most of the students think ought to be the governing idea of truth, especially for this place. This is because society expects the university itself to be a place that tries to make us more and more deeply aware of the world; this means helping us understand what is really there and what is not. Students would be surprised and even dismayed if their professor was deliberately saying things that had no verisimilitude. Students and faculty think of this as a place where one tries to represent the world as truthfully and accurately as possible.

As the lecture progresses, the professor is now arguing that a reason that places are important for understanding projects is that they provide an indispensable means by which we create rules about what should or should not happen. The lecture hall itself, she might say, has multiple sets of rules, some implicit and customary, others based on decrees from the university and from statutes and laws of the municipality and the state. She may even point out that an important time-honored expectation is that a lecture hall should be devoted to the pursuit and instruction of truth. But in the context of place, truth needs the support of two other virtues: justice and the natural, and that this is the case for this lecture hall too.

Now suppose the professor is in midsentence about this very last point when the door to the hall opens and a young man in jeans and backpack bursts through, heading toward the professor and blurting out that he has been robbed, and could the professor tell him where to find campus security.

Several things may be racing through the instructor's mind. She may think that since this person looks like a student, and so would be someone who would understand what is supposed to take place here, an intrusion like this would occur only if he had in fact been robbed. He is then what he claims to be—a victim. She may further think that, though the purpose of the place is to discuss intellectual issues, rectifying an injustice and caring for a person in need can temporarily trump the pursuit of truth. This is what she concludes and so stops in midsentence and directs him to campus security, wishing him well.

Pausing in the lecture to care for someone is a manifestation of treating people justly. It involves the belief that we have certain duties and obligations to strangers. We should try to help them if we can. Indeed, the teacher finds this student's need so compelling that for the moment, her attempt to dispense care and justice takes over the place's focus on the search for truth. But as soon as the student is helped, truth again becomes the dominant thread.

The student's entry illustrates an unusual problem of justice for this place. Normally, in this context other strands of this virtue are present and play supporting roles. Such strands derive from various sources, including the university's rules about who may attend such lectures. These rules involve criteria for admission, perhaps including grade-point averages and Scholastic Aptitude Test scores, and Advanced Placement examinations. It also includes departmental requirements such as meeting of certain prerequisites. These are explicit rules. But there are de-facto rules that impinge on justice. The class may not have many low income or minority students because the cost of admissions is high and the financial aid package is too small to offset it. These are issues the university wishes to rectify, but how? It can raise more money for fellowships, but should it also make an effort to admit on the basis of economic need? Other issues of justice may bear more directly on the conduct of the class. The professor is expected to assign grades on the basis of merit and not on some sort of personal preference. Need, which is an important quality of justice and is often linked to mercy, is not relevant here. No matter how much a student pleads, saying "I need an A," the professor should not relent. And equality, is also not the issue. She is not supposed to assign everyone the same grade, for that is not fair to those who have merited higher ones.

These issues of justice are all very important, but in this case they are there to support the pursuit of a correspondence theory of truth. If the admissions standards are to change, if grading is to be assigned by different criteria, then the justification would have to be that students would learn more about the world—that the place would better serve its avowed function of helping faculty and students (and the community at large) to pursue the truth.

We have been discussing the dominant role of a particular conception of truth in the classroom and the way some strands of justice do or do not support it. But it is important to recognize that the room also contains elements of the natural. We find the natural in two forms: as a world external to us, and as part of the biological world within us. The solar system, with its forces and energies, and a good deal of the natural processes of the earth, are part of this external nature; our internal nature refers to the fact that we are physical and biological beings undergoing processes beyond our control.

In the academic classroom it may be mostly (though not entirely) internal nature that comes to mind as the realm of the natural. The students and the professor are biological beings. They digest, breath, and even age as they attend a lecture. They cannot help this. And one among them may not be able to help the fact that he has been overwhelmed by a high fever and an excruciating headache. This flu prevents the student from concentrating. Suppose that this student, in the middle of the lecture, walks up to the professor, tells her about the symptoms and asks permission to leave. It is likely that the teacher may excuse the student because he is in the grip of something that he cannot control. It is simply natural that he cannot focus. An element of nature then becomes an issue that for this moment and in this place appears as a reason that someone behaves a certain way. Honoring this force of nature, or giving into it, becomes a virtue or the right thing to do, and, under these circumstances in this particular place, the natural temporarily trumps truth and justice.

External nature is also part of each and every place, including the classroom. All places include air, gravity, heat, and the like. Many of these can be modulated in the classroom in order to support the project taking place. So, heating, lighting, and sound are controlled to allow for the greatest degree of communication and concentration. Suppose that the lecture takes place in the dead of winter on a campus in the upper midwestern part of the United States, and that the heating unit fails to function. Even though the windows are closed, the temperature in the room begins to fall. Soon the room becomes so cold that it is perfectly natural that students and instructor become uncomfortable and are no longer able to concentrate. Or, suppose that this classroom were struck by a tornado, shattering windows and causing damage to the roof. The class would disband as people sought shelter. The reaction to these forces of nature is that it is natural (not just or true) that we submit to them; they are beyond our control. The natural itself becomes a justification for our own actions. So, in this place for the moment, the natural again trumps truth. But, this is after all a classroom, and when it is open again, it is truth that is supposed to be the primary virtue, with justice and nature playing supporting roles.

Other strands of truth can be found in lecture halls. Right next door to this one may be a lecture on the philosophy of mathematics. There, the professor presents the position that the kind of truth that animates work in pure mathematics is a coherence theory in which truth must be internally consistent whether or not it points to anything real and outside of itself.[5] Across the hall, in a sociology class, a consensual theory of truth is discussed. Here the idea is that a member of a community tends to believe things to be true that most if not all of its members agree are true. Having the truth based on such a consensus reduces conflict and increases social cohesion and identity. And, in a course on religion, a professor may discuss a revelatory theory of truth, which claims that what is true is revealed to us by God.

In each case, though the strands of truth being discussed are different, two kinds of claims are likely being made by the lecturers. First, that this or that kind of truth is practiced in one or another context—e.g., pure mathematicians seem to follow a coherence theory when they do pure math, and members of a social community who want to preserve social cohesion may be driven by a consensual theory of truth. The second is the more general claim that these statements about how truth is being understood and used in a particular context are themselves offered as true statements or models of these situations, and thus meet the criteria of a correspondence theory of truth. These other kinds of truths then become points of view that can be discussed and taught as part of the university's general commitment to a correspondence theory. Being committed to a correspondence theory of truth does not lead in practice to a consensus about the world. On the contrary, there are many competing views about what the world is like. The only thing a correspondence theory commits one to is the assumption that the validity of a view depends ultimately on its degree of verisimilitude, and that a view that seems best may be so only provisionally, for there may soon be another that is closer to the real.[6]

This idea of truth may describe the practice within most universities in the West, but not all. Some have overwhelming religious commitments so that the purpose of the institutions is to better communicate the revelations of God. Insofar as this revelatory theory supercedes a correspondence theory, such institutions are suspect by the academic community at large, and may not even be thought of as true universities. These points about different criteria of truth being linked to different places and contexts are exactly what both instrumental and intrinsic geographic judgments would expect to be the case, but instrumental judgments will be content to view them as a claim that truth is situated and relative to particular contexts, while intrinsic judgments would not.

Another layer of issues about truth (that will be developed later) can be illustrated by the content that is being taught within classrooms. In the 19th century United States, universities held to a correspondence theory of truth as they do in the 20th century, but the content of that truth was different. In many of these places one would find racial theories being offered and accepted as accurate pictures of reality, in that races were believed to vary in mental and moral capacities. And for most in Nazi Germany, including those within the university, racism remained the factual model of reality. Even in the "hard" physical sciences, models about the world change, so what is taken to be true at one time is not quite the case in another. The point then is not only do different places claim to emphasize different types of truth, but even for places that agree about the kind of truth they wish to portray, the content of that truth differs from place to place and time to time. Such complex relationships of truth to context is also what both instrumental and intrinsic judgments expect, but again interpret them differently, as we shall see.

Places then emphasize different criteria of truth, and a single individual, as he or she moves from one to another, also is expected to adhere to different ones. When she is a scientist in a laboratory or a classroom, she may believe in a correspondence theory. She may also be party to a consensual theory of truth when involved in faculty meetings, and she may also attend religious services that she believes reveal something about divine providence. Each of us can thus experience different strands of truth as we move from place to place, and places that expose us to new ones go beyond schools, universities, and churches, to museums, libraries, galleries, and studios, to monuments and places of memory.

What types of place stress issues of justice as their primary concern (Fig. 4)?

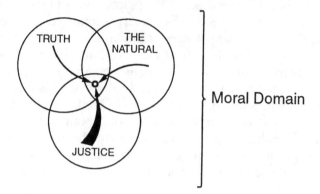

Fig. 4

Justice

Some places claim to focus on justice. This is true of the entire legal system. A court of law applies principles of justice. Lawyers, judges, and juries mete out justice, and litigation is supposed to be a means of redressing injustice. Prisons, as places of punishment and rehabilitation, are part of the criminal justice system. Legislative assemblies such as state houses, congresses, and houses of parliament develop and promulgate concepts and rules of justice. Other places focus on more particular issues of justice and care. Charitable organizations, soup kitchens, halfway houses, homeless shelters, community centers, and the like, focus on various qualities of justice that mostly emphasize need. Even nation-states, in their support of equality, liberty, and fraternity, are claiming to be places enforcing and advocating specific qualities of justice; and democracies are supposed to be more just than other political forms.

Justice also turns out to be the dominant virtue in a great many places in our everyday landscape that do not advertise themselves as standing for any moral principle in particular. We can tell because, as soon as claims are made that such places are not doing what they were supposed to do, these claims are usually framed by appealing to qualities of justice. Almost all places of work fall into this category; in fact all places do, when they are seen primarily as sites of employment. When we think of an office or a factory, one of the first things that comes to mind is what they are involved in producing. But in a market economy, these places and their products are means of making money. From the perspective of both the owners and the workers, a central set of issues is the generation and distribution of wealth and the conditions of employment. Workers in a factory perform tasks that when challenged soon invoke issues of justice. The factory may be violating contracts. It may be endangering the worker. It may not be paying workers enough for their labor. Unions may enter the debate, claiming that the employer is exploiting the workers, while management may claim that workers are not loyal or responsive. Confrontations can lead to mediation, arbitration, or to picketing, lockouts, and strikes. These acts are buttressed by appeals to justice and fairness. Certainly, claims about truth and the natural are part of the assumptions, but it is the justice of these relations that is the issue. The dominance of justice in the workplace arises most vividly in a market economy when alternative forms of organizing work are proposed. Worker syndicates, co-ops, worker-controlled organizations, and communes are proposed precisely because their principles of organization are thought by their advocates to be more just than those of private ownership.

Qualities of justice emerge whenever a place is seen primarily as a site of employment; this is so even for places that are explicitly involved with dispensing justice. The many workers in the courthouse—the clerks, janitors, secretaries, and guards—who also may see their work in terms of promoting this goal of justice, are also employees of the court system and their own labor conditions raise issues of fairness that may have little to do with the substance of court decisions. As a verdict in a courtroom is being rendered, these employees may agree that the judgment was just, but are still meeting to discuss the injustices in their conditions of employment. Or, the volunteer in the soup kitchen may believe that he is providing care to the needy, but may himself feel that other volunteers with whom he works are not fairly allocating tasks among themselves. Similarly, there could well be at least one person on the line who sees the justice of donating food, but who believes that the soup kitchen itself is a symbol of a failed system of social justice—a system that makes it unlikely that he can have a home and job.

Even the university faculty, who claim to be searching for the truth, may become embroiled in conflicts with the board of trustees over salary and working conditions. They may decide to boycott classes, to strike, or form a union. At such times, issues of justice trump those of truth; but unlike other places of employment, the primacy of justice here is temporary, and even as issues of justice surface, these will be guided by the faculty's more basic quest for truth, unless of course, the nature of the place is changed so radically that truth is no longer that important.

The Natural

These same complexities and dynamics attend places that are driven by the virtues of the natural (Fig. 5). Natural processes are a part of every

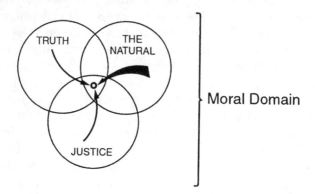

Fig. 5

place, but we tend to think of places along a continuum of intensity and degree of natural forces. At one end are urban environments that are primarily artifacts of culture. Here we believe that human decisions are the most important in determining what takes place. At the other end are wilderness areas. These are places that we set aside so that the predominant forces controlling what takes place are the natural.

Of course, no place can be purely natural or purely cultural, for in creating place, we are weaving these together to varying degrees. So places that focus on the cultural such as cities, factories, and classrooms still contain elements of the natural, and are often under its grip. Similarly, a wilderness area by definition contains aspects of culture. We, as cultural beings, to some degree are controlling aspects of what takes place, even if that means only preventing some things from being there. We limit and control human activities there, and even often prevent specific elements of nature from migrating in or out. Should a wilderness area that once had grizzly bears have them reintroduced? If so, then it must be recognized that the presence of the grizzlies will affect other creatures and habitats in the place: and an effort must be made to keep the grizzlies within the borders of the park.

There can never be places of pure nature. Still what takes place can be thought of as mostly natural when arrayed along the nature/culture continuum. And an important reason that nature is set aside as a place is that the natural itself is seen by many as a virtue. People enter wilderness areas to escape the "impurities" of culture, to find the wholesomeness and harmony of nature, to commune with nature, and to become one with it. In these senses, nature is a source of something good, and our distance from it, a source of problems. But, as we said, even this type of nature—nature as wilderness—requires culture. Beyond the fact that we maintain the boundaries, there are as yet no such wilderness places that exclude humans from visiting. How could they if part of the attraction of nature is to be surrounded by and immersed in it? People by the thousands visit Yosemite, Yellowstone, and other wilderness areas. They enter through roads, park in parking lots, camp at campsites, and walk on trails. And these places are maintained and controlled by numerous park employees. So even though few would doubt that the forces of nature are more direct and visible here than in cities, we still find issues of justice arising because of how visitors are treated and how the government handles its park employees. And truth also plays a part, as when debates occur as to whether the habitat that exists now is really the one that existed before human occupancy, or, more generally, to what degree the landscape is really "natural."

There are also numerous places that emphasize different strands of internal nature. A hospital tries to restore one's biological nature, and a mental hospital one's psychological nature. And each of these too is interwoven with different elements of truth and justice. Again, most of the places we mentioned may not advertise themselves as encouraging this or that virtue, but when they are contested, their emphases become clearer and debates about place ultimately revolve around the mix of these virtues. There is a point at which the change in mix can make a place entirely different. This can occur even in a place that does not specialize in its mixes of virtues, but rather attempts overall to balance the three. A readily accessible example of how the mixes are flexible, but only up to a point, is the home (Fig. 6).

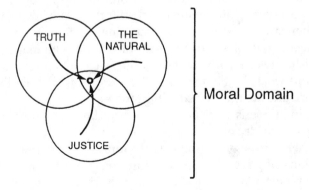

Fig. 6

A Blend of Virtues: The Home

Homes come in all shapes and sizes. A home can be the hearth, or collection of hearths in a hunting and gathering band, or the hut or village in a peasant agriculturalist society. For someone in an industrial society it can be an apartment, the community, or even the nation-state. Although varied in form, home is found in practically all cultures. In all of these cases the idea of home suggests a valued ideal, even when the homes in actuality may be living hells. As a moral ideal the home is supposed to be a nurturing place relatively unspecialized and fluid in its mix of virtues which should be largely under the control of those who live there. Again, this is the ideal, and very often not the actuality.

Consider a father, mother, sister, and brother living in an apartment in some North American city. In daily life, the family is constantly engaged in drawing upon the three virtues and their numerous strands. For example, the parents love both children equally and wish to give them equal attention. But one child has a learning disability and needs help with his homework, while the other child does not. Doing what is fair or just may be to spend more time with the child who has the greater need. Here, need trumps other strands of virtue. But when it comes to celebrating a holiday, the parents provide both children with gifts that are of equal value. Now, justice as equality becomes important. If one child is constantly disobeying the parents, breaking household rules, while the other child does not, the parents may reward the obedient child only, in which case they are dispensing some form of justice on the basis of merit.

Justice in any of its forms is not any more important in the family than caring about biological needs. Issues of health and cleanliness, of food, rest, and exercise take up a good deal of time and space (for example, the kitchen, bathroom, and bedroom are mostly devoted to the body), and become intertwined with the threads of justice. And of course there are issues that involve teaching children about the world, about meanings, and about the realm of truth. Children are taught to distinguish between truth and falsehood, fact and fiction, and to try to understand what the family believes the world is like. Truth, justice, and the natural are intermixed within the walls of the house.

Balancing these is what takes place in ordinary circumstances. But even though the idea of home is capacious, it may be the case that when one of these virtues takes hold for a long time the home becomes another kind of place. If the children are consumed by intellectual pursuits, the family may then devote most of its time, space, and attention to issues of the mind. They may invest in a large number of books, use the dining room table to do homework and as a place to discuss intellectual matters, and read books and plays out loud in the evening. The place may become like a school or even a university classroom. Or, if one of the children develops a serious and chronic ailment, all of the energies then are devoted to restoring the child's health, and the home becomes like a hospital. But if a child is out of control socially, then discipline and rules become the dominant issue. From the viewpoint of the parents, the house is now a battle zone. From the child's view it may be a prison.

The mix of virtues within a home can shift from moment to moment. The range depends in part on the mixes in other places to which the home is connected. These also affect the content and interpretation of the virtues. The threads of justice within an Islamic home in Afghanistan under the Taliban, may range among need, equality, and merit, but their interpretation and application will be different here than, say, in a middle-class home in North America. In Afghanistan, attention and love may be given to both male and female children equally, but when it comes to satisfying intellectual curiosity and encouraging precocity, the opportunities between the sexes may be very unequal. Male children may be permitted a far wider range of options than female children. In contemporary North America, a conscious effort might be made to provide both males and females with equal opportunities.

The loomlike qualities of place make things interdependent because place helps weave together and affects what takes place, including moral judgment. Both instrumental and intrinsic judgments recognize this. But instrumental judgments alone are not able to go farther than to say how these mixes are changing and contingent. Before we pursue this issue further, the structure or loomlike quality of place needs further explanation on two counts. First we must consider how it is that place is concerned with these three virtues, and second, how exactly does place help weave them. That is, how does the structure or loom really work? The first question brings us to the connection between the moral and the empirical.

The Empirical Domain: The Realms of Nature, Meaning, and Social Relations

The virtues of truth, justice, and the natural are part of the moral domain that place engages. Why it is the case that place helps weave these three can be understood better when we consider that the moral is related to the empirical, and here too place is an essential part of the connection. Roughly speaking, the three virtues—truth, justice, and the natural—correspond to, or are the moral equivalents of, the three parts or realms of the empirical domain. These are meaning, social relations, and nature (Fig. 7). The point I want to stress here is that mean-

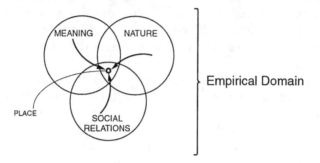

Empirical Domain

Fig. 7

ing is associated with truth, social relations with justice, and nature with the natural (as represented in Fig. 1).

Meaning, nature, and social relations are the empirical level at which place operates. (Meaning and social relations constitute culture, and nature is the rest of reality. Or, to put this another way, the dual relationship of nature and culture becomes a trifold one when culture is divided into meaning and social relations. This trifold relationship corresponds to how most complex cultures view the realms of reality.) Elements from each of these realms were implied in the section above in each and every example and description of the virtues. Instead of saying that a university classroom focuses on some element of truth, we could have said that it explores the meaning of a given topic. Meaning here includes our ideas, values, and beliefs, as well as how we project these onto the world through language and other symbolic systems. In her discussion, the professor includes only some ideas or meanings of place and geography and may even have created a new one. In another class the topic could have been the meaning of the electron. But if one asks why it is that she is pursuing this meaning rather than that, and whether ultimately it was the right one, we are again back into the moral realm of truth: the professor saw these meanings or concepts and ideas as the right ones because they most closely corresponded to reality— they led to the truth. Overall, disputes about meaning then are resolved by appeals to truth—whether it be correspondence, coherence, consensual, or yet another type. Meaning is animated by truth.

What about justice and social relations? When students at the university demonstrated against the Vietnam War by conducting sit-ins and teach-ins, or, when a later generation of teaching assistants protested against working conditions and salaries through strikes and pickets, or when faculty protested against their working conditions by forming a union, they were raising issues about social relations. They were discussing the relationship between Americans and Vietnamese, between teachers and students, between employees and employers. But if we ask why the students or faculty did what they did, their answers would most likely contain appeals to particular conceptions of justice. Justice, and perceived injustices, are animating the concerns about particular social relations. And, the moral import we give to the natural as something beyond our control animates our reactions to nature. In the case of the classroom example it is not that a virus or a tornado are virtuous, but rather that it is virtuous to recognize that there are things that may be beyond our control. It puts our lives in perspective.

We can of course discuss specializations among places without mentioning the mixes of their virtues or the moral level at all, but rather refer only to their mixes of empirical qualities—their weave of meaning, social relations, and nature. So, to return to the places mentioned above, universities and museums are all places that claim to focus on aspects of meaning. Indeed, they are sites that generate or preserve meaning. They also contain social relations and elements of nature, but these are woven together to support the different strands of meaning. Prisons, law courts, and state houses are places that enforce, rectify, or even create new types of social relations (an illegal alien for example is a social class or group defined by law), while the vast majority of offices, factories, and work places in general, through their focus on selling products, making money, and providing wages, are all emphasizing (and creating) social relationships between employees and employers, between income levels and economic classes, between producers and consumers, or between employed and unemployed. These places do contain meaning and nature, but rather it is social relations that predominate. And of course parks and wilderness areas are places that emphasize elements of nature. They too contain elements of social relations and meaning, but in supporting roles. And when the loom is used to unweave and reweave the virtues, it is also doing so for the strands from the empirical realms (indeed, weaving one set also weaves the others).

Again, it is important to note, as we did in our discussion of the virtues, that places are not only about one thing. Rather I am saying that many places are specialized, which means the things that take place are supposed to support the primary or ultimate project for which the place is set aside. This also means that such purposes can and should be questioned. And, when they are, then a new mix or weave may replace the old. And even a specialized place that is not challenged can have other things taking place that do not have to be linked to its purpose: e.g., a research laboratory may also be the place where two people fall in love.

A key difference between the instrumental and intrinsic is this: the instrumentalists see the mix of the moral and its qualities to result from the mix of the empirical. The virtues of truth, justice, and the natural are driven or constructed by the elements of meaning, nature, and social relations (as indicated by the direction of the down arrow)—and many social theorists see social relations as the predominant force, hence their claim that moral issues are "socially" constructed (Fig. 8).

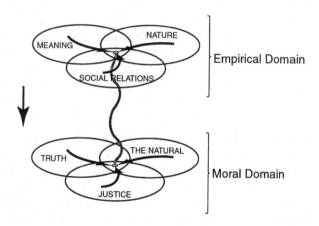

Fig. 8

But those committed to intrinsic judgments would turn the argument around. While they would recognize that the empirical has a bearing on the moral (and that moral evaluations must be sensitive to the most subtle empirical nuances), the empirical does not produce the moral. The reason the moral is not reducible to the empirical, and the reason it can in fact animate the empirical, is that the good is real and is not a product of the rest of reality. This allows our intimation of the good to direct these mixes of empirical and moral qualities (as indicated by the direction of the up arrow), and to rescue them from relativism (Fig. 9).

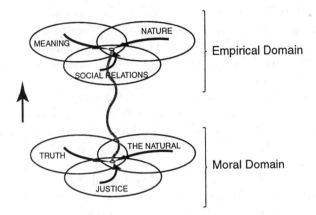

Fig. 9

Since I wish here to explore what it is like to be in a world that is instrumentally driven and that argues for the relativity of the virtues, I will proceed for the time being with the instrumental assumption that the moral is constructed by the empirical. This is the view we will take for the sake of argument, and it also means that we can illustrate the way place's loomlike structure works by focusing on the empirical, which, according to instrumental judgments, would be the factors determining the moral.

According to this view, what we think is just could stem ultimately from the realms of nature—as in our biological nature. This is a claim that is made by evolutionary psychologists and sociobiologists when they argue that morality itself plays an evolutionary function and can be explained by the laws of natural selection and by our genetic makeup. For instance, the desire to care for others could be explained, and in fact predicted, by the need to protect and even sacrifice one's life for those who in turn can help the long-run viability of one's genetic pool. Or it could be the realm of social relations that provides the causes for our sense of morality. For a capitalist system, Marxists would argue, a sense of justice focusing on individual rights and freedoms would be expected, not because these are intrinsically good, but because these values can be given particular meanings that can then function as support for a host of necessary attitudes for the development of capitalism, including fostering an ethos of consumption (which makes the personal accumulation of material objects and wealth the means of expressing individual rights and freedom) and the avoidance of class consciousness and identity (which would seem irrelevant to a "free" individual who believes consumption is the path to personal gratification and self-fulfillment). Or the realm of meaning can be argued to become the principal generator of morality. Freud, for example, saw moral concerns as embodied in a component of our mental makeup called the superego, which in turn represses other elements of our psyche such as the libidinal forces and the ego. Morality then is a way of regulating otherwise socially disruptive psychological forces.

These are simply instances of arguments readily at hand claiming to be able to reduce (in the sense of subsume or explain away) the virtues (or the realm of the moral) to empirical relations and forces. These arguments make the moral appear to be consequences, by-products, or "epiphenomena," of natural, social, or mental relations. The moral is not animating these realms, but rather a result of them. This fits the instrumentalist view that sees what is right or wrong to be something arising from, and justifying, particular projects. Hence, the mix and weave of virtues in a place would depend on the mix and weave of empirical elements. As these change, so too would the moral. The virtues of a place could thus be "read off" from the mix of empirical elements.

Even if this reductionist position is correct so that the moral is relative and dependent on the empirical, we should still take a closer look at why there are three realms. Why does place weave meaning, nature, and social relations? One part of the explanation is to consider first that these three—as a "trialectic"—stem from a duality that pervades what it means to be a geographical being. I refer here to the observation that we are both part of and apart from nature. We arise from material form, are subject to evolution and biological laws, and yet also seem apart from them. In terms of the problematic, place-making is one means of understanding this dualism. Each place combines elements of both. And so from the beginning our place-making allows us to be both part of and apart from nature. This dualism moves easily into a when we think of the cultural as being composed of social relations and meaning.

To show how simple this is, consider again, the dualism. First we have our natural side, often described in terms of the body and our drives and instincts. Yet we think we are able to rise above these because of the capacities of our mind or reason to resist them. So our mental and intellectual qualities set us apart from our biological or natural qualities. This division between mind and body, as it is often called, is not yet equivalent to that of nature and culture, for the mind is a more specific category than culture. What it leaves out are social forces and relations which are also part of the category culture. So, as selves, we possess a natural part, an intellectual part, and a social part. Each alone has been used to characterize us at one time or another: we have been called natural or biological beings, intellectual beings, and social beings. Place makes us aware that we are all of these.

We create places and are the ones who use them as tools, but part of their function is to help us draw together the threads of our own being. The role of place in interthreading nature, meaning, and social relations affects the way the self connects these three realms to form our own identity (Fig. 10). Indeed, when we say that places help us un-

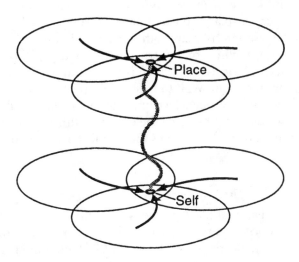

Fig. 10

dertake projects by drawing together these three realms, we are also saying that place makes it possible to form selves. The project of making the self, like the other projects we undertake, requires place. I do not mean the simple-minded idea that a single mix in one place magically determines the mix in oneself. Rather it is that numerous places constituting our environment help form the mixes of nature, meanings, and social relations that become part of ourselves.

Place helps empower us, and we in turn create, sustain, and transform place. We would not be able to combine in ourselves the elements of nature, meaning, and social relations without the help of place, and of course place cannot exist and combine these elements without the agency of the self (Fig. 11). Seeing ourselves as geographic beings makes

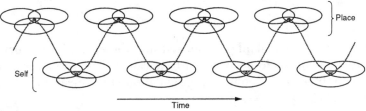

Fig. 11

us aware that we are all three, and that place enables us to tie them together through our projects and the role place plays in our identities. Self and place then are intimately interthreaded: a theme that will be taken up again.

I say that place helps tie these realms together, or to draw elements from each and weave them. But how can place help us do something? As with languages, places make things possible because they have effects. And these effects may both enable and constrain us in our tasks. We know something of the structure of language, of its grammatical and syntactical forms. What then is the structure of place? Specifically how does the loom work to help us weave? Since we are looking at the problem only through instrumental judgments, I will illustrate the working of the loom by using the empirical threads rather than the virtues, for instrumentalism argues the moral can be read off from the empirical.

The Loom: Its Three Components

The loomlike quality of place provides the central set of structural relations that all places possess and that allow them to function and affect ourselves and our projects. The image of a loom (and looms come in a variety of forms) is useful in discussing how place works. Place as a loom contains three interrelated components (Fig. 12): the rules of

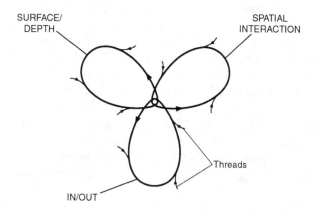

Fig. 12

in/out of place, spatial interaction, and surface/depth. These structural components of the loom (which I have called elsewhere a causal circuit with three causal loops) engage and weave the threads (again indicated by the small curved arrows).

The central feature of these components and their interconnection rests on the idea that in order for elements of one realm to affect elements of another, they must be in contact through space. Space is a necessary form of connection. The three components provide this connection because each is a variant of spatial effects that conforms to the way space is used in a particular realm. In/out of place fits within social power. Spatial interaction is compatible with the laws of nature. And surface/depth engages meaning and awareness. While each is suited to a particular realm and thus can be thought of as ranging through it, each, as will be explained shortly, can engage elements of the other two realms in the overlapping areas (Fig. 13). This is how

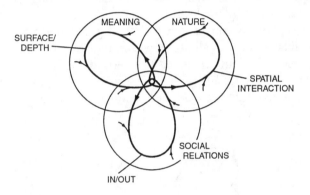

Fig. 13

place has an effect and how it functions as a loom. Each component is equally important, so it does not matter where we begin.

I will start arbitrarily with the *in/out of place rules*.[7] All places (in the sense we are using this term) possess rules about what should or should not be contained within the boundaries (Fig. 14). This is clear

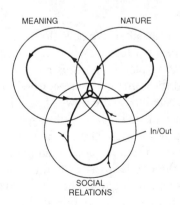

Fig. 14

simply from the logic of clearing a place and keeping it cleared for certain things to occur. These rules may be explicit (as in laws) or implicit (as in customs, and everyday practices), and they may apply directly to the place or be part of the rules of a larger system which includes this place. Let us leave aside the issue of whose rules, or for that matter whose place, and consider that whoever is in it and using it will find that there are rules. The classroom can illustrate all of these points. It has a host of in/out of place rules designed specifically for it. The room is set aside so that a particular subject will be taught by a particular professor, and will be attended by students who have registered for that class. The rules may also be more implicit or customary and cover the courtesies that members of the class owe one another. And the rules can be set by the local state, which may prohibit smoking in public space, and of course by the federal government which asserts that federal laws apply to all places within the nation.

These rules of in/out are linked to a second component, and that is the flows through space and within the place—or the *spatial interactions* that occur both within the place and among places (Fig. 15). The flows

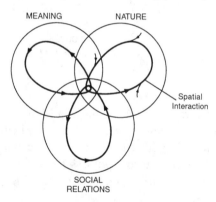

Fig. 15

and the rules affect one another. We have flows in part because we have rules, and we have rules because of what we anticipate to be the flows (both within and among places). So, this room has rules about attendance, and also about the control of temperature and humidity, because of what might be expected to be the flows of people, moisture, and temperature in the absence of such rules. And by having these rules, people who might have been somewhere else are here; and the flows of energy in the natural realm too are being contained or deflected by the walls, windows, roofs, and the regulations of the internal environment. A nation-state has rules about immigration because of what its citizens think might be the flow of people to that nation in the absence of such rules. And again, these rules affect the flows. Those who wanted to come may not be able to, and now must stay or go elsewhere to other places with other rules. Changing the rules changes the spatial flows, and changing the flows will necessitate changing the rules.

These two—in/out of place rules, and spatial flows and interactions—are coupled to a third—*surface/depth*—which exists because the combination of rules and flows partly weaves together a landscape, an appearance, or a surface within the place (Fig. 16). Even a place as small

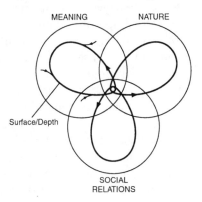

Fig. 16

as a room has such qualities. What is taking place in a lecture on geography has the appearance of a classroom—students sitting and taking notes, a teacher standing and lecturing. Larger places also have appearances or landscapes. The appearance or surface of any place can be called into question. Even though the classroom looks like a place of learning, what may really be taking place is an exercise in social power. A national park has a look to it or an appearance, but is it really natural? The surface or appearance then can be challenged in terms of not being what it claims to be. Rather it is "merely" an appearance or surface, disguising what is truly taking place beneath or behind. In other words, when we realize that through rules and flows we are weaving a landscape, the meaning of this weave can be called into question—it can be problematized.

Surface/depth, in/out, and spatial interaction are interconnected. Activate one and you will likely activate the others. A change in rules will change the flows, which will change the appearance of the weave. A challenge to the weave, by saying that it is only that—an appearance, a front, or a stage, and that the real is being obscured in the "backstage"—can lead then to a change in rules, which will change the flows, and so on. But what constitutes the flows, the appearances, and the things being controlled by in/out rules of place? They are the elements of meaning, nature, and social relations (and also the virtues of truth, justice, and the natural)—the elements we said are woven together in every place. So these three components or properties of the loom—in/out, spatial interaction, and appearance and reality—are the means by which the particular threads or elements are engaged, woven together, and then unraveled, and new ones taken up.

Here is how these structural parts actually help us draw things together. As the figures have indicated, each is embedded in a particular realm and engages elements from that realm and from the overlapping areas of the other realms. The rules of in/out help us include or exclude elements of all three realms—of meaning, nature, and social relations (see Fig. 14), but as they do so, the elements tend to become socialized. That is, in/out rules themselves are a type of force that we find in the social realm. A meaning that violates an in/out rule is socially inappropriate: if the professor were now to utter profanities in the classroom, these meanings would violate a host of rules and not be socially acceptable *here* in the class, but they may be acceptable in another place. The rules also socialize nature. Rain is accepted out there in the forest or garden, but when the rain enters the room it becomes a leak. As soon as the leak is repaired and the rules about water reestablished, then it is rain again, because it is outside. So too with the soil that is trekked in on one's shoes. This is dirt, but as soon as we sweep it out beyond the threshold, it once more becomes soil.

When the spatial interaction component is the mechanism that engages elements of nature, meaning, and social relations, it naturalizes them (see Fig. 15). Rain, and the flow of water, and the movements of particles and masses and the waves of energy are easily thought of this way, which reflects the fact that space and its properties such as distance are fundamental elements of the natural world. But the spatial interaction component also does this to social relations. Focusing on the flows of migrants (or of any other group) from one place to another makes them appear as bodies in space. It draws attention to their material qualities such as the numbers, their location, and even the velocity and volume of the movements through space.[8] And, in the same way, the spatial interaction component naturalizes elements of meaning in that it emphasizes their thinglike properties such as velocity and direction of movement that can be modeled in terms of bits of information flowing through space. A map showing the origins and destinations of this book from the publisher to the store to purchasers would be an example of the way the meaning represented on these pages becomes conceived as flows through space. Another example would be a map of radiating lines from a single point indicating the daily circulation of a newspaper. The density of these lines reflects the flow and interaction of meanings through space.

Finally, when the appearance and reality or surface/depth part of the loom engages elements, it tends to *problematize* their meanings (see Fig. 16). The Yellowstone National Park in the United States is supposed to be a monument to nature, but as a place, it also adds elements of social relations and meaning, and the resulting landscape—the things and connections that take place there—can be challenged as inauthentic. Is this really nature or natural, or simply one group's conception of it? And if so, then what rules and flows must change to make the appearance more like what it should be? And is a university lecture hall really focusing on what it claims to be its purpose—the realm of meaning—or is it simply an appearance that disguises the real process of conveying power? And if so, then what change in rules and flows would we need to be able to have the place more closely coincide with what it should or professes to be?

So the rules of in and out, the spatial flows, and the surface and depth are the three mechanisms of the loom that every place possesses to engage and weave together these elements. A most important point is that these are woven together without reducing one of the elements to the others. This means that in the absence of place, the elements of the realms become impossible to connect in fact; and in theory, we can think about their connections only by having one of the realms subsume the others, or translate the others into its own language. Roughly speaking, natural sciences focus on nature, social sciences on social relations, and branches of the humanities on meaning. To bring these separate realms together conceptually requires translating or reducing one to the language, terms, or theories of the other. So the natural sciences can reduce meaning to mental states and brain waves; the social sciences can reduce nature to a resource or to a social construction; and the humanities can reduce science and social science to languages. This is what happens conceptually when we think nongeographically; and without real places these realms can never actually meet in a sustained way. But recognizing the role of place in everyday life shows that although it weaves them in varying proportions, still, it does not translate or reduce. So here in this room all three realms are participating, and though the realm of meaning should be more important, it does not explain away or reduce the others. The air we are breathing in this room is not simply a symbol, or a resource, it is also a molecular structure that affects our biological state. The reason place can combine and weave without reduction is, as we said before, that all three components are variants of the same relation—space. Each simply reworks space so that it can engage the material of a realm. In/out makes the spatial into rules and power, spatial relations makes it adaptable to the laws of nature, and surface and depth makes it part of meaning and awareness. The three components then make place far more complex than space as a causal agent, and central to human projects.

The components are always in effect for each place, but they are usually invisible when the place is functioning according to plan. We become aware of them when a place is dysfunctional, or when its purpose is being challenged. Then one or another becomes activated. It does not matter in theory, though it is important in practice, which one is activated first: any questioning of the meaning of a place (surface/depth), or attempts at rearranging its rules (in/out), or altering the flows to and from it (spatial interactions), will have the potential of changing the dynamics among the components, and of altering the elements that are woven together.

I have illustrated the loomlike structure of place—its causal circuit and loops—through examples at the empirical level, because I am focusing here on instrumental judgments and these see the moral as a product of the empirical. But I could have discussed the dynamics of place by using the virtues in the moral level. That is, the same components engage elements of truth, justice, and the natural as they engage social relations, meaning, and nature. Again, it is a key point of interpretation between instrumental and intrinsic judgments as to whether the moral elements are woven as a result of the empirical, or vice versa. As I have said before, I believe the moral can be independent of the empirical and animate it, rather than be a product of it, but this issue will be the focus of attention in chapters on intrinsic judgments. Here I will continue to take the instrumental side and illustrate the loomlike quality of place through the empirical examples, turning to the moral in the next chapter.

It is this capacity to weave without reducing that makes place necessary for undertaking projects. It is important to remember that we are focusing on place, but the same analysis applies to projects. The projects too are mixes or weaves of these three, because they are undertaken in, and are products of, place. How can one be trained as a physicist without being in a place that produces such knowledge and educates such people? How can a domestic cow exist without the support of a place called a farm? How does a hybridized form of maize arise, without the impress of the place in which it was cultivated?

The Weavers

This then is the loomlike quality that places provides. It allows us to weave and unweave. (There can be of course multiple individuals or groups trying to weave, and in this sense we can think of each attempting to control the loom or set up their own loom.) Those doing the weaving do not even have to be in the area. This is true for wilderness areas, where the rules, flows, and meanings of the place are established by voters, lobbyists, and politicians, who do not reside within its boundaries. The place then is controlled remotely and by people in a variety of locations. Similarly, a multinational corporation controls the working conditions in factories around the world—the CEO, the board of directors, and the stockholders are the ones deciding the patterns and pulling most of the strings, although unlike a wilderness area, people live and work within these weaves. If we focus on who pulls the strings, we may be able to draw on specific models and theories about the relationships of power and powerlessness. These may suggest that certain distributions of power lead to certain kinds of weaves or patterns. For example, gender theory may have us expect that patriarchic society will weave places in expected ways, and Marxist theory might alert us to patterns produced by capitalism (and both theories would anticipate that the geographical surfaces would be used to disguise what is really taking place). These can certainly add to our understanding of the generation of the patterns, but still, these concepts of gender and class, or more generally, these theories of power (and of who does what to whom) themselves do not explain why place is important to control. The idea of place-as-a-loomlike-structure provides an understanding that fits within any theory of power. What is more, there is no guarantee that theories of power are correct. In that case, the model of place itself can be used inductively to trace the lines of power. That is, we can examine a place, or part of one, and see at a moment in time how a particular thread was pulled, who or what group decided and were empowered to make a certain rule, who challenged that rule or a particular weave or appearance, who else entered the discussion, and what the consequent rules, flows, and challenges to appearance occurred and how this process led to changes in the place's fabric. Unraveling the weave along the lines of the loom can then lead inductively to the actual power relations operating in particular contexts. This is why the focus of our interest is on the loom, and not directly on those doing the weaving.

I would like now to turn to the weavers, but not yet as parts of groups identified by one or another social theory, but rather in the general terms suggested by the model itself. That is, I would like to address some of the general ways in which human beings and this model of place are related. The most direct way of discussing this is to reconsider some of the key relationships between place and self.

Place and Self

Place of course is created and sustained by us as agents, and we need its effects as a tool. The primary point we have addressed is how place helps us to undertake projects. Here I would like to draw attention to a number of ways in which place helps to identify and affect us. The general reason why place can be a form of self or group definition lies in the continuities between the mixes of elements of nature, meaning, and social relations that constitute ourselves and that constitute the places that influence us (as indicated in Figs. 10 and 11). If, for example, one is intellectually inclined (that is, his or her personality stresses the realm of meaning over the other two), then one might feel at home in a place such as a university. But another person there whose primary interests are in the social could feel very much out of place. These mixes and matches can be more finely tuned according to the type of intellectual interests and the offerings of the institutions. None of these relationships are static. The mixes within our own selves change, and so too do those we find in place. The geographic problematic makes it axiomatic that we continuously find ourselves out of sync with our landscapes: our own mix of elements and those of the places can never completely correspond.

Rarely do we consider place and identity in such a fine-grained manner, though the structure of place could suggest the means of doing so. Rather we discuss the connections in coarser terms. Scale may be the factor that is singled out. In ascending geographical scale, we can imagine a person's identity being linked to a village, a city, or a state. Ethnic groups often feel attachment to a cultural core area. In these cases, there is a complex and often mystical link between person and place, so that past events and even the place's soil seem to course through the person; his or her identity "stems" from this place. The fusion is intense for some cultures like the Australian Aborigines who believe that their bodies or selves are simply a manifestation of the life force that emanates from an ancestral territory. So strong is this connection thought to be that they would find it difficult to occupy the territory of another group, for that would be like trading ones identity for another's. While few modern cultures see identity and place to be this intertwined, some forms of nationalism, irredentism, and ethnocentrism push us in this direction. Nazi propaganda tried to fuse German character with German landscape through its concepts of blood, soil, and *Lebensraum*. Other nation-states attempt to fashion links between identity and place, though not to this degree. In doing so, a claim is being made, perhaps implicitly, that a necessary, essential, and stable quality of one's identity comes from place. These attempts can all be thought of as a way of thinking of self and place as somehow essential, and of place as somehow thick with meaning and power. Indeed, there are people who attempt to forge this link and can be thought of as professional "place-thickeners."

In many respects, this need to fuse and create a sense of a rooted and stable identity is a reaction to the contemporary condition of high geographical mobility, rapid development of specialized places, and their constantly changing and contested nature. These have the effect of thinning out the meaning and significance of place. When the places that provide us context are seen to be contingent and constantly changing, then so too would our identities. Understanding that the projects we undertake in each place leads us to different functions that we must perform, and hence to different roles that we must play, moves us in the direction of being identified with these place-defined roles. At home, one is a son or daughter, at school, a pupil, in the playground a member of a peer group, and the particular meanings of these roles vary depending on the matrix of places to which they are linked. As adults, we hold even more and varied roles defined by places of work, travel, and leisure. And in the consumer's world where advertising tells us that we can become what we consume, including the landscapes and places that we shop in, vacation at, and move to, our identities may be immensely fluid and multiple. But the same fragmentation and fluidity can cause as many mismatches as matches. We may not yet have a place to be what we truly are, or we may be forced to assume an identity we do not want. These issues of mismatching and identity are not only personally frustrating but lead to serious problems of prejudice and injustice. One may be gay in a homophobic society; a single and poor mother in a welfare-less community; an intellectually precocious child in an academically indifferent school district. Such people are marginalized and in substandard places; and they too must have a place that enables them. These, and countless other examples, as the expression goes, is the difference place makes.

The place not only enables and constrains in terms of its mix of elements, but in its literal and figurative spatial relations with other places. One may be at the core or heart of things, or on the margin.

This is true at any scale. A business can have an inner sanctum, a front and back stage, a privileged place where decisions are made and other places that execute them. A city has numerous places of privilege—a financial center, an elite neighborhood, a posh shopping center—that set off and influences the rest. And nations too can be in the core or the periphery, or the "first," "second," or "third" world, or be a bridge between them, or change its "position" under new circumstances. And those occupying these places, especially those who are most involved and empowered by them, become identified as central and core, or peripheral and marginalized.

Place has these effects because of its components. While all three of the loops work together to make place into a tool, each itself can become a dominant factor in molding identity. Consider first how spatial interaction can be a defining quality. Commuters move through space—they are part of the spatial flows, which is then being used to define them. Nomads even more so, as well as vagabonds and Gypsies. Of course, such people are still involved in places as they move—the commuter in an automobile or in a railroad station and on the train, and the nomad in tents or yurts pitched in particular grazing territories under the group's control—but they are identified primarily as part of flows and spatial interactions.

The in/out of place rules define individuals and groups not in terms of their movements, but in terms of whether or not they belong. They can be in place or out of place, or have no place because they have been displaced. People can be punished by being exiled or banished from place, or they can become homeless or refugees. And then the surface and depth component can be part of one's definition, as when groups or individuals occupying a place are called interlopers.

The starkest connection between self and place is revealed as soon as we look directly at the fact that a person cannot *be* in the sense of *exist* without a place. If we decide that the homeless must not be here, and never say where they are allowed to be, but persistently say not here, we are then in effect erasing them from existing.[9] People cannot exist without a place, but they cannot exist if the place itself completely disempowers them. Place should enable individuals which means at the minimum that the individual(s) in it must to some extent control its structure and dynamics. An extreme case of one that allows no control is that of solitary confinement, where the person confined is not in touch with changes in nature, has no contact with others, and has no intellectual stimulation. This place will lead inevitably to the disintegration of the self, because the individual has absolutely no control over its structure and dynamics.

GEOGRAPHIC IMPLICATIONS

Place then is not simply a passive cipher or linguistic trope marking the spatial coincidence of elements of meaning, nature, and social relations. If it were, it could then be replaced by (or reduced to) just these elements. Rather it is an indispensable means by which these elements are not only occurring together but also interacting without having one reduced to the other. Another way of summarizing the significance of the loomlike quality of place is to see what it says about contemporary geographical themes. I can do this best by briefly discussing its implications for particular geographic topics that fall into two groups. The first are those for which instrumental and intrinsic judgments come to the same conclusion. The second are those for which the instrumental is insufficient because of its emphasis on situatedness, self-interest, and a lack of concern with free will. The second does not yet introduce the intrinsic, but points to the necessity of something like it to supplement the instrumental.

Implications Compatible with Instrumental and Intrinsic Judgments

Dynamic landscapes   The model is about the dynamics of place within a system of places in space. There is no perfect place, and its boundaries and mix of elements from the empirical and moral realms will change. We will never stop transforming the world, and never stop creating and changing place.

The social construction of space and spatiality   To say that space is socially constructed is misleading for several reasons. Unless we are using the word "space" completely metaphorically, we must take our cue about space from the physical sciences, which tells us that though it is conceptualized and modeled by human beings, physical space is still thought to refer to something out there that is not made up by us. If the social construction of space is to have any meaning, it would refer to the places and the flow among them that we construct. Furthermore, to say they are socially constructed privileges social relations. To avoid this it would be better to say humanly constructed, which leaves open the issue of which realm dominates. Extending these points means that instead of "spatiality" we should use the term "platialility" to discuss the mutually constitutive roles of geography and the self.[10]

Space of flows   Modern culture seems to make places more like nodes in a network and the interconnections among them into "a space of flows," to use Castell's terms.[11] Now these terms may capture a quality of being in the modern, or hypermodern, world, but they do not redo the nature of place and space. The structure and dynamics of place assume that places are interconnected and explain how the elements and loops generate spatial flows and interactions (Fig. 17). (Some

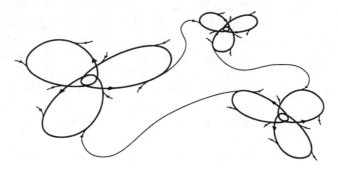

Fig. 17

places, such as transportation terminals, are designed explicitly to facilitate flows.)

Scale　Place is a human artifact and so too is scale. The structure and dynamics of place are independent of metrical size, and they are also independent of their place in a spatial hierarchy. They are scale-independent in both senses. Scale in both senses—of size and place in a nested and functional hierarchy—is derivative of the structure and dynamics of place, and especially those in/out of place rules that stipulate a place's size and its functional relationship to other places. These sets of rules along with the other parts of the loom, generate the features we mean by scale and their uses as a tool. These make sense of the fact that the scale of place, just like place in general, can be used as a strategy to resist, or contest, or circumvent, or obfuscate.[12] One kind of scale-related property though deserves attention because it is so often used. I am referring here to what appears as the inescapable fact that no matter what kinds of places we create, if there are more than a few, there will tend to develop the need for a two-part distinction in functions of place, which translates into a rule-based relationship between two geographical scales or levels. One level, such as a house, a school, a factory, consists of those specializing in particular projects. The other level, such as an administrative unit or a political territory, will be those that focus on coordinating the relationships among the first that derive directly from the geographical proximity of these places, even when these places may have nothing else in common but proximity. Once the places of the second level are formed, they affect the relationships among the first in new ways. Still, the functions of the two levels or scales are rule dependent. This relationship of two levels is generally found in most systems, and each of the two can have numerous sublevels. (This general property of scale will be discussed in Chapter 7.)

Thin, thick, and rooted   Places that are specialized tend to be thought of as thinner in their mix of elements than those that are not. Non-specialized or thick places would be common in a hunting-gathering community in which so many facets of life—meals, education, politics, and entertainment—take place together around the campfire, or in the tepee or hut. We then would tend to see such places as layered with these qualities and think of them as thick, relative to a place of learning like a classroom, where the mix is designed specifically for one function. After this function is completed, and the students and teachers leave for the day or the year, the place is virtually empty, for no intended function is taking place. A thick place is often thought of as rooted to its particular site and its dense weave or tapestry appears to have emerged from its location. Often now in a world that is uprooted in which so many of us are displaced or constantly moving through space and occupying thinned out and mass-produced places, being rooted in a specific location seems warm, nurturing, and secure and so there are many who intentionally try to thicken place, the ultimate thickening mechanism being ritual, magic, and the sacred. But being in such places is not necessarily a virtue for they can be exclusionary and narrowing, fostering an essentialist identity.

Context   As place helps us weave, it creates a context. Moreover, this idea of context-as-place allows us to speak of things being contextualized without making the process relative. This is because many places attempt to create a similar or identical weave, and so produce for the most part a standard context with a predictable effect or product; and also because when the place's mix of elements is singular or unique, which may in fact be the intention, the mix still results from a process of weaving that is common to all places.

Hybridizing nature   Once we bound nature and make it into a place, it immediately becomes interthreaded with social relations and meaning. Nature and culture become hybridized. True, each type of place specializes in the mix and emphasis, which affects how we then experience nature, meaning, and social relations in our everyday experiences, but without place these could not be interconnected.[13]

Change, transgression, contestation The loom that place provides can not only weave but unweave, so it is also a tool for contestation and change. Who operates the loom? The model does not tell us (for that is up to other kinds of theories), but rather argues that whoever is in charge will have to use this loom, and whoever challenges those in charge will also have to use it. A challenge to power can be initiated through any of the loom's components. For example, in/out rules can be transgressed.[14] These transgressions may be slight and playful, as in various forms of graffiti, or the transgression can be more serious, as in the sit-in demonstrations of the American Civil Rights movement, or the sit-ins and teach-ins during the Vietnam War.

Or the challenge can be applied directly to the flows and interactions, as when a picket-line is formed or a boycott is enacted to stop the spatial flows. Or the challenge can be leveled at the meaning or symbolism of place—as when demonstrators claim that what is said to be taking place is a sham, a stage, and that the real and more important events are not being given voice. Again, affecting one triggers the others, and places change. The point is that to change place we must use the same mechanisms that allow place to be constructed, but in the other direction.

Implications Incompatible with Instrumental Judgments

Place is the key instrument in the geographic problematic, and the structure and dynamics of place explains why. They point out how place is an indispensable tool that enables, constrains, and often defines us. It also points out how place can help weave together nature, social relations, and meaning even within our own selves. Both instrumental and intrinsic judgments accept this, but they part company at the point where the self becomes an autonomous agent. Because the instrumental sees our judgments as ultimately caused by and rationalizations for our conditions and interests, it does not leave room for us as autonomous agents who, though influenced by nature, meaning, and social relations, and by the places that interthread them, are not simply their instruments, but rather can and ought to remove ourselves to some degree from their grip. Instrumental judgments do not leave room for selves who can decide, even when it is only to an infinitely small degree, but still decide what to do, and thus not be only and entirely instruments of these forces. This recognition of free will shifts the problematic from an empirical study to one that is also moral. Still, without yet developing intrinsic judgments any farther than I have in the introduction, I will use several contemporary issues to illustrate how instrumental judgments cannot address contemporary issues that most need addressing, and thus, by implication, suggest that something more is needed—namely, intrinsic judgments.

Situatedness Situatedness is the general claim of instrumental judgments that we are products of the elements of nature, meaning, and social relations woven at particular places—that our behavior and our selves are caused by these (even though somehow we may have initiated them). The causes can come from one big place, as in claims that a nation-state determines the behavior and values of its citizens; it can come from a generic set of places containing general sets of forces, as in the claim that particular economic conditions and social relations, such as class, that can be found scattered around the globe, in the last instance determine our behavior and our beliefs and make those affected by these forces behave and think alike; or the causes can come from numerous and constantly changing places, as when we see the places of work, home, nationality, recreation, and a host of others, providing shifting bundles that determine who we are and what we do, and thus make us all different. Whether it is one localized set of forces making us alike, or multiple and shifting ones making us different, the central argument of situatedness is that we are products of these forces. Arguments for situatedness, as we shall see, are part of most social science views, and also of postmodernism and a good deal of Marxism.

The problem of situatedness is its inconsistency. True, we are always in places and affected by them. Yet even as we are, we must be able to think ourselves out of them simply to be able to manipulate them and be aware that places are indeed affecting us. Thinking ourselves out of them and achieving a critical intellectual distance means we are making an effort to exercise some degree of free will (which is the point of intrinsic judgments), but free will is a concept that geography has never really addressed directly.[15]

Essentialism Essentialism is the attribution to individuals and groups of essential and unalterable characteristics and identities due to their surroundings. In geography, essentialism comes about when there is argued to be some essential and unalterable relationship between people or groups, and the characteristics of place, even when the self and the places with which it is connected are numerous, dynamic, and fluid. Essentialism arises when we do not assume there is free will: without free will we are victims of contexts (whether a single, stable place, or multiple and fluid ones), and what we think we ought to do would itself be a product of them.

But free will does not mean that all essentialism is gone. Rather it means that *the* essential quality to being human is such a will and its accompanying attributes of reason and imagination. In other words, it postulates an essential self as a morally responsible decision maker with a will.[16] (This is why the self, and not the body, becomes the central and essential moral concept.) Intrinsic judgments would also postulate the need for place in order for the self to be empowered, but denies that place can determine us. All of the rest may well be contingent.

Relativism   We have seen that place can be contested and challenged, and that surfaces can be reduced to mere appearances. But what happens when the concept of reality itself is challenged—a challenge that instrumental judgment is not equipped to fight and might even condone—so that there may be nothing real to hide and thus disclose, only another surface. We then enter a condition of skepticism that currently is expressed most deeply in the views of postmodernism, and in this case through one of two particularly important strands; namely the Derridian or deconstructive strand that claims we can never escape language, and so virtually everything, and everything geographic, is a form of language.[17] A geographic place becomes, in Derridian terms, something like a "discursive site."

A second strand in the postmodern view draws on Foucault.[18] Foucault allows that there is more to the world than language, but that we are still prisoners, and this time prisoners of contexts and forces that surround us and make us irretrievably situated and pawns of power. It is these power relations that determine the accepted and dominant meanings and systems of representations, for these are products of and serve the interests of such forces.

The central problem with postmodernism is that it does not provide a means of articulating a positive view of what to do. This may be justified for a field such as literary criticism that is concerned with representation, but it is not for a field that undertakes to examine the kinds of places we have created and asks additionally what places we should create. Geography demands to know the positive: What should we do? What is a good place, and what is a bad one? We must have answers because we must have places. To not have places is not to have selves, and to make places without knowing if they are good is to run the grave risk of making the world worse than it is.

PLACE GUIDED BY INTRINSIC JUDGMENTS

Those who agree with my arguments so far may still voice skepticism about the need for geography to ask what the good might be. Could this problem of the good not be left to others? Certainly others should address it, but I do not believe it should be left only to them. The point of this book is that geography can illuminate aspects and relationships of the good that would not otherwise be as visible. This is the general reason for exploring the topic. Another, though minimalist, reason is, if we don't

try then we are supporting instrumentalism by default. This arises, paradoxically, from geography's success in dealing only with the instrumental.
Understanding the power of place—its structure and dynamics—deepens geography's capacity to make us think that the moral is produced
by the empirical conditions in particular places. That is, geography can
readily contribute to understanding the "apparent" situated nature of
moral views. I call this apparent because it appears so only if we are not
aware of the possibility of intrinsic judgment to guide it. And if we do
not pursue this possibility of intrinsic judgments—or pursue geography's approach to the good—then, whatever we may personally believe,
the apparent becomes real by default. So, to reveal again and again only
that moral claims are linked to particular contexts leads by default to the
view that they are products of these contexts. Avoiding this default position requires that we focus on the possibility of articulating what moral
qualities can animate the "ought" in the geographic problematic. For
that, we must turn to intrinsic judgments and the role of the good.

How does the picture of the power of place change with intrinsic
judgements? It would of course include everything we have said so far,
though with modifications and additions. But, at this point, a sketch
would not adequately render these details. This is so for two reasons.
One is that geographers have spent most of their time on the instrumental. The intrinsic is far newer terrain, and its qualities will require
the rest of the book to articulate. A second and more fundamental
problem is that intrinsic judgments are not as easily rendered in visual
form—as lines on a page—because these judgments concern the most
abstract ideas about the real and the good. Still, at the risk of having
our drawings be taken too literally, and having the real and the good
be made too concrete, a fuller picture emerges if we have Figs. 18 and

Instrumental Judgments

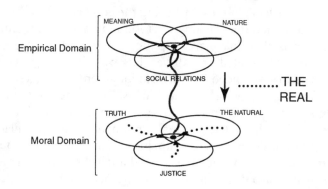

Fig. 18

19 appropriate the core of Figs. 8 and 9. The instrumental picture (Fig. 18) emphasizes the claim that the moral qualities woven in place are a result, or product, of its empirical qualities. The instrumental has the moral reduced to the empirical. In addition, it adds reality as a whole (indicated by the real on the right side)—that infinitely receding horizon that makes itself known only partially and provisionally. Of course the place itself and its empirical and moral qualities are part of, are affected by, and alter reality. The reality on the right side is intended to refer to all of this and most importantly to an awareness of it. Even though reality is ultimately ineffable, the dotted line indicates that for instrumental judgments, the degree of awareness is especially tenuous because it is driven by self-interest and our situatedness.

The intrinsic drawing (Fig. 19) reverses the arrow to indicate that the moral animates the empirical and is not a product of it. The draw-

Intrinsic Judgments

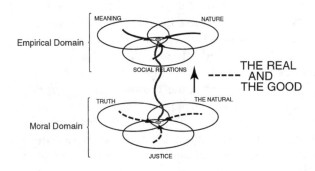

Fig. 19

ing adds not only the real but the good as part of reality to indicate that intrinsic judgments encourage us to create places that increase our awareness of reality and the variety and complexity of that reality. The drawing indicates this, and the fact that our connection to the real and the good is less tenuous, by having the good added to the real and by having a dashed line. Having intrinsic judgments guide our place-making means also that places are more open and transparent, and that we are less situated and driven less by self-interest. This more openness of places guided by intrinsic judgments is suggested by the lighter shading for places in the intrinsic drawing, and the darker shading for the instrumental drawing. Also, the drawings suggest another important difference. Intrinsic judgments encourage places specializing in different weaves of truth, justice, and the natural, but at the same time its emphasis on altruism and altruistic gift-giving for each of these realms avoids having these virtues become relative. Retaining variety and complexity among the virtues without succumbing to moral relativism is reflected in the dashed lines for truth, justice, and the natural in the intrinsic drawing.

This may be as far as our diagrams can take us. It is does not seem feasible at this point to visually render how intrinsic judgments lead to a greater openness of place; or how places can be used to increase our awareness and increase variety and complexity; or how intrinsic judgments create a system of dynamic checks and balances so that places are neither too transparent nor too opaque; or how these criteria encourage a system of altruism that is manifested in different but complementary ways through truth, justice, and the natural, and how this system is the means by which moral relativism is avoided; or how both the capacity of free will and self-deception operate.

Perhaps the most basic obstacle to representation is this: How can we convey the reality and ineffability of the good without mistakenly making it appear as something visible and material? This important assumption that the good is real and ineffable may seem strange, but only if we assume that the only real things are material things and relations (things that are some how more tangible) that fit into a material world. This is of course a position which many are inclined to hold. Those who do would likely challenge not only the sense of the good I am putting forth, they would also challenge other ideas, such as the one that the mind is really something different from the material qualities of the brain. That is, they would challenge the idea that the sense of consciousness and mental states that we experience when we have thoughts and thoughts about thoughts are truly real. They would claim instead that these mental states may feel real, but they are only by-products of these real electrochemical reactions. Certainly, this challenge—this reduction—is worth serious attention, but it may well be wrong; it certainly feels wrong, and many have argued it is wrong. Those who do are thereby arguing for another layer to the real. They are claiming that consciousness along with free will are real, but different from the reality of gravity, atoms, molecules, and electricity. My point here is that to make the claim then that the good exists as part of the real is to make a claim just like these other nonreductive claims. It does not lead to anything supernatural or religious. It simply asserts what seems to be obvious—that reality has multiple facets, some of which are more or less accessible, and not all of which can be known in the exact same way.

I

Instrumental Judgments

3

SITUATEDNESS AND RELATIVISM

THE OVERARCHING POINTS OF THIS CHAPTER and the next are that instrumental judgments promote moral relativism and moral absolutism, and that moral relativism cannot be a defense against absolutism, and can turn into it. How this is so can be summarized briefly this way.

If I am a relativist, and you want to impose your view on me, I have no grounds to say you shouldn't, for your view would be as important as mine—just different. But if you do impose yours on me you then are acting as an absolutist and I, as a relativist, am allowing and even abetting it.

Since relativism does not deter absolutism and can even encourage it, this chapter is about relativism. It explores how the situatedness of instrumentalism is part of relativism. Simply put, an instrumental position argues morality to be a product of, and a rationalization for, particular positions and self-interests. Hence, moral claims are relative to these. Exploring moral theories that make such claims or come close to doing so is the object of this chapter.

Place's loomlike quality can explain how it helps twist virtues such as truth, justice, and the natural to conform to our self-interest. So convincing are these twists that it may come to seem that there is no escape: self-interest makes all moral judgments circular and relative. Let us consider those philosophies that seem to be explicitly and positively claiming an instrumental position, and thus are rejecting the possibility of intrinsic judgments.

Theories such as these differ in what they think the significant forces to be. Sociobiologists would see the realm of nature as playing the determinate role. Other theories would have the dominant factors lie in the social realm (class, gender, or simply social "power") and still others in the mental or the realm of meaning. I am not after the details of these arguments, but rather their shared and general idea that our moral positions are results of and instruments for these forces.

I will start with the simplest kind of claim, and go on to more complex ones found in communitarianism and postmodernism. Even though these two support situatedness, they show some signs of dissatisfaction with it. In postmodernism the situatedness usually refers to the unique character of different places. I will end the chapter with a consideration of some themes in Marxism. The situated and relativistic strand of Marxism comes from its general emphasis on the unfolding power of social relations (especially class) and the capacity of this power to mold our ideas and values—to make them products of this situatedness (which, as opposed to postmodernism, is a situatedness that can be repeated in countless places)—and on Marx's pronouncements that morality is simply a rationalization for this class interest or situation. I want to juxtapose this Marxist theme with another strand from Marxism, its appeal to what might be taken as an intimation of the good, and how the tension between the two cannot be overcome without having Marxism become less of a social and more of a moral theory, a point which will be taken up again in Chapter 7.

FROM EMOTIVISM TO COMMUNITARIANISM

At the smallest scale are those views of morality that claim it to be nothing more than a means of rationalizing what is important to particular individuals at the moment. If I enjoyed walking in a nearby forest that is being threatened by developers, I may join a group that hopes to preserve the forest by having the land transferred to a state-controlled park. Justifying my position, I may say that this transference of land will lead to natural conservation, which is morally right. What I really mean is only that I want very much to preserve this area because I value the natural. Claiming this is morally right is simply a way of saying natural conservation is important to *me*, or that *I* prefer it.

An *emotive* theory of morality is the philosophical claim that not only do people use moral terms this way, but that this is probably the only way they are, or can be, used. According to Alasdair MacIntyre, "emotivism is the doctrine that all evaluative judgments and more specifically all moral judgments are nothing but expressions of preference, expressions of attitude or feeling, insofar as they are moral or evaluative in character."[1] Since these preferences and feelings are caused by forces surrounding us, we and our moral judgments simply become agents of these place-specific forces. The use of truth, justice, and the natural become defined by context, is a rationalization of context, and is relative to context. Emotivism is instrumental.

A closely related view is that morality is simply what is customary to, or practiced by, a group in a place. Morality is equivalent to *mores* or customs. As with emotivism, the moral is being used as a justification for what is practiced and preferred, but now it is a group, a collectivity and the place they occupy, rather than a single individual, making this claim and reinforcing it for others. The moral conventions of a context or a place act as a force and stipulate what should be done.

Variants of emotivism and morality as mores or custom can apply to larger-scale places and contexts, and fill in the intermediate range of our instrumental continuum; these arguments become more complex and subtly change when we consider the position of *communitarianism* at the other end of the continuum.[2] This moral position (or collection of positions) argues that what is good is not simply what is motivating us or what is customary, but rather finds the good as an ideal embodied in the proper and best performance of particular roles and social relations.

Communitarianism draws upon the Aristotelian conception of a moral life, which extends the older Homeric view that sees moral action as behaving in a way appropriate to one's function in society: being a good and moral father, host, or warrior means performing admirably what these roles require. Aristotle abstracted and amalgamated these role-specific functional definitions so that they became appropriate to the conduct of an entire human life. Humans have specific needs and goals by virtue of their constitution, and being good means mastering those virtues that lead us to—that are instrumental for—our peculiarly human purpose, or *telos*. Because these virtues are means to this most important end, they constitute living a good life. But for Aristotle, these purposes are not possible to conceptualize outside of, or beyond, the life of roles, duties, and obligations in particular communities. And for him, there was only one good place or context for a human life to be lived, and that was the city-state. (One cannot be fully human outside of this context.) The human being should then follow the purpose of a human life, which in the broadest and best sense, is defined by the duties and obligations of a citizen in a Greek city-state.

Communitarians, too, consider what is good and bad in terms of one's duties and obligations in the context of living a human life, but this time in the modern world. A fundamental difficulty in adapting the Aristotelian view to contemporary life is that expectations placed on individuals within a modern community are shifting and contestable. Indeed, Aristotle was confident about his statements concerning human purpose and the virtues required to live a moral life because they were reinforced by the relatively cohesive set of social practices of the Greek city-state (although one can argue about exactly how cohesive Athens was in Aristotle's time). But as modern culture becomes more dynamic and its places more fragmented and yet integrated in a global system, meanings and social contexts are no longer stable and self-reinforcing.

A communitarian such as MacIntyre would like to retain the Aristotelian view of the moral, but modern conditions have forced him to rephrase the position so that it no longer stresses human life and *telos* in general, but focuses instead once again on human practices. His is still a more abstract position than the Homeric, for practices are more general than roles, which can be included as practices. A practice is "any coherent and complex form of socially established cooperative human activity through which goods internal to that form of activity are realized in the course of trying to achieve those standards of excellence which are appropriate to, and partially definitive of, that form of activity, with the result that human powers to achieve excellence, and human conceptions of the ends and goods involved, are systematically extended."[3]

Because of the dynamic—and what MacIntyre would argue the immoral—qualities of modern life, the kernels of good practices reside in our smaller day-to-day activities, in our games, our work, and our roles in family and occupations. At this level people can achieve competence and control over their actions and thus exercise virtue. "A virtue is an acquired human quality the possession and exercise of which tends to enable us to achieve those goods which are internal to practices and the lack of which effectively prevents us from achieving any such good."[4] Virtues then according to MacIntyre, though not necessarily organized around the three—of truth, justice, and the natural—which arise from a geographical focus, nevertheless still are products of and embedded within practices, projects, and places.

Reliance on the Aristotelian idea of projects embedded within contexts is characteristic of other communitarian thinkers. Michael Walzer, for example, uses "spheres" of justice rather than practices and projects as the primary focal point. A sphere is a rather general conception identifying a social nexus of practices within which particular concepts of justice pertain. Within these spheres, "different social goods ought to be distributed for different reasons, in accordance with different procedures, by different agents; and all of these differences derive from different understandings of the social goods themselves—the inevitable product of historical and cultural pluralism."[5] The family, the school, the workplace, and the polity are such spheres. Each tends to focus on particular types of social goods and the appropriate or just way they are distributed. For example, in modern Western culture, free exchange may be the criterion for distributing goods in the marketplace; in the school it may be in terms of merit or dessert; while in the home one may focus on a range of criteria. The critical point is that "all distributions are just or unjust relative to the social meanings or the goods at stake."[6] While Walzer does not focus on place but rather on a set of spheres and their meanings of justice, his argument at this point is similar to how the weave of the virtues and their significance depends on place (and my arguments about variety and complexity and the evils of tyranny draw upon his). In the classroom, for example, the most important strand of justice would be expected to be different from that in the soup kitchen; but unlike Walzer's analysis, our focus on place allows us to consider the apparent situatedness of other virtues as when the strand of the natural in a wilderness area differs from that in a hospital.

These meanings as well as the spheres develop historically. Since practices change, and their boundaries are fluid, only careful observation of practices within particular spheres in actual cultures reveals the appropriate principle of justice. If principles in one sphere spill over and dominate another, then this is unjust for the dominated sphere. It is a form of *tyranny*. If, for example, the market place, which may be the appropriate model for the sphere of the economy, becomes the model for the political, or the educational, then we lose control over that portion of our life.

MacIntyre, Walzer, and other communitarians focus on practice. The good is in and defined by these. There is little "place" for a nonrelativistic good. The idea that there might be, they argue, comes from abstracting us too far from the particular situated cases and from the conditions of being human. Indeed, much of communitarian motivation is a response to what they see as the limitations and dangers of the more abstract approach. For Bernard Williams: "The belief that you can look critically at all your dispositions from the outside, from the point of view of the universe, assumes that you could understand your own and other people's dispositions from that point of view without tacitly taking for granted a picture of the world more locally familiar than any that would be available from there; but neither the psychology nor the history of ethical reflection gives much reason to believe that the theoretical reasonings of the cool hour can do without a sense of the moral shape of the world, of the kind given in the everyday dispositions."[7]

For communitarianism, the abstract and the concrete do not mix well. Indeed, pursuing the abstract runs the risk of our no longer being in touch with the contexts and particularities that make us individuals with moral interests. "How can I that has taken on the perspective of impartiality be left with enough identity to live a life that respects its own interests?"[8]

The way they handle these concerns constrain communitarians from moving farther along the path to a more universal view. Their preferred position is situated, or in Walzer's terms, "radically particularistic"—"One way to begin the philosophical enterprise—perhaps the original way—is to walk out of the cave, leave the city, climb the mountain, fashion for oneself (what could never be fashioned for ordinary men and women) an objective and universal standpoint. Then one describes the terrain of everyday life from far away, so that it loses its particular contours and takes on a general shape."[9] In contrast to this, Walzer means "to stand in the cave, in the city, on the ground."[10] For him, as for other communitarians, morality is situated, contextual, place-dependent, and thus relative.

In many respects, the instrumental uses of the structure and dynamics of place described in Chapter 2 can help inform the communitarian ideas of the individual practices of excellences in MacIntyre's view, or the spheres of justice in Walzer's. In this vein, a good society is one in which the instrumental values of the place are allowed to be played out mostly on their own, independently of other places. This relative isolation allows the weave of empirical elements to produce a weave of virtues that become accepted by the weight of tradition. This process can lead to a richer set of contexts, a greater set of options for living life. But on its own, instrumentalism or communitarianism cannot then address the questions: Are these good? How do we tell? Communitarianism does not develop a good that is independent of these contexts. It leaves too much room for the possibility that a local practice and particular realm of life, or a sphere of justice in Walzer's terms, may in fact be an oppressive realm, one which is different from the others, but which exploits or subverts or tyrannizes its own members. Because of communitarianism's premises, there is little it can say that helps identify such cases or that provides a means of redressing them. Again, the problem is that the particular practices, spheres, or places generate their own purposes, *telos*, or goals, which are then to be evaluated on their own terms, and not according to some more abstract set of principles.

Even though these philosophies stress situatedness and shun the idea of an independent good, they nonetheless seem to assume something like it. For example, in order for the assumption that people should be allowed to develop according to their own inclinations to make sense, it must rest on the deeper and more universal claim that people deserve autonomy and respect. Or the idea that such developments will create a variety of practices, which in itself is good, must rest on a more universal value of variety and complexity. Most communitarians, though advocating a form of situatedness, are not happy with relativism, but have little ammunition to fire against it. There is little in MacIntyre's position that decides what is a good or a bad practice, still he feels compelled to offer something as an alternative to relativism. He does so in his not very clear notion that the reflexivity of a tradition and its ability to confront crises and inconsistencies will somehow lead away from relativism's excesses, while Walzer has suggested that there may be a "kind of minimal and universal moral code"[11] to which we can appeal. But again, if sharing values and respecting the rights of people who share values to develop their own projects is important, is it not because we assume more generally that human beings have rights and duties that go beyond the particular? And if Walzer thinks such shared positions that may comprise the foundation of his own theory are so "thin" as to be virtually devoid of meaning, then how does he expect the details of communitarianism to have moral weight for others than himself and those who come from his context? The problem with communitarianism is that it does not have the abstract reach necessary to expand upon these claims. It works well if the world is already working well. The variety and complexity quality that they stress is not counterbalanced by seeing to the real.

To have this abstract reach, one's view must become less situated. This cannot come about by making every place alike. That would be a form of tyranny, and would play further into the instrumentalist position, for it would have all views become products of just one type of place. Rather to have this reach requires the exercise of our will and reason to consider how our different positions are not determinate of our views of the good, but rather opportunities to see it. The problem remains that any form of instrumentalism is stunted and unrealistic for it leaves out a most important part of reality—the good.

POSTMODERNISM AND THE REALITY OF THE GOOD

Communitarianism values the situatedness and relativity of moral claims. Postmodernism (where I again am attending to only a few but important strands from this complex movement, and extending them to our concerns about morality and place) seems to argue that morality is not only situated at all levels, from the individual to the group, but that this applies also to the claims of communitarianism. What

I want to develop here is the highly speculative thesis that one can also detect a certain dissatisfaction with this situatedness. What I mean here is that while postmodernists make a strong argument for relativism and situatedness, they appear to hold out hope that making this claim will somehow help liberate us from such conditions. I will develop this thesis by examining the connections between two ideas that can be discerned in some postmodernist positions. Again, I must say that there is no single and consistent set of views that characterizes this "ism," but the Foucauldian and Derridian are important ones, and it is their puzzling moral implications I will discuss.

It will be recalled (from the end of Chapter 2) that the Foucauldian strand stresses situatedness and power. Knowledge is situated and is a product of the power of the situation. The wielders of power are in turn subject to sources of that power. They are its instruments. There is little or no possibility of becoming less situated and more detached. I do not want to go into detail here about the contradictions involved in holding this position[12]; rather I will draw attention to its possible and perhaps unintended moral implications. I see these as arising from a puzzle that develops when one considers the motivation behind postmodern work.

The puzzle begins to unfold as soon as we recognize that if a claim is made, presumably by a nonpostmodernist, that we are or can be less situated or more detached, the postmodernist would see this claim itself as a situated one promoting a specific set of power relations and interests—namely, the one in power and making the claims. The postmodernist would argue that these claimants are justifying their position by calling themselves less situated and more clear-sighted. And the postmodernist would say that self-interest applies equally to the communitarians, for they too can be promoting their own positions through their moral theory. Indeed, the postmodernists would claim that all theories, whatever their particulars, are products of power, and hence relative to the context. There is then little likelihood of knowing if one position is better than another.

The puzzle becomes clearer if the question is asked: Why then does the postmodernist make any claims at all? Why is he or she not silent? The highly speculative answer I would like to entertain, and which some postmodern literature may imply (and the answer that Foucault gave, even though he eschewed the possibility of developing a moral theory) is that to disclose the links of power to knowledge is important to do because it is morally emancipatory. It will free us.[13] I do not mean here only the claim that disclosing situatedness will free us from absolutism (which though a hope of postmodernism, is not attainable through relativism, since relativism may actually encourage absolutism) but an even more radical proposal that disclosing it will some how allow our true selves to emerge. I will return to this issue after we consider the second postmodern strand—the deconstruction of Derrida (though Derrida eschews the postmodern label).

Here we find another set of turns, but this time deriving from the role of language. All knowledge, as this strand emphasizes, is expressed as and through language, and so we can never go beyond the symbolic system. If there is a world out there, we cannot know it except as a form of representation. There is no escape from language and its circularity and inconsistency. The meanings of terms may not be anyone's in particular, but float around, like the entries in a giant and continuously changing dictionary. Everything is seen and experienced as a language or a text. If statements in a text refer to something outside, as in the idea that the word tree refers to a "real" tree out there in nature, a postmodernist would point out that a real tree is itself a part of a text or set of meanings that belong to another language, perhaps that of biology, and that "out there in nature," also refers to a text, perhaps that of the natural sciences in which biology is a part, and so we can never get outside of the textual quality of experience. Statements about reality, or about the nature of things, are simply statements in the context of a text. The important point here is not that humans use language or symbolic systems, but rather the claim that they can never get outside these systems and know if they refer to anything else but other languages and symbols. The position holds that it is impossible to make the distinction between the linguistic and the nonlinguistic. This is so even for place.

Applying this to science means that science, as a symbolic system, is not making statements that correspond to, or model, reality, even approximately. Rather the statements of science are but "texts" about another text described as "reality." Evidence, proof, verification, or disconfirmation and other apparent points of contact between the language of science and the real world are actually points of linguistic connections between scientific statements and statements in related languages that encompass sets of meanings and symbolic manipulations that we translate as, or take to be, "observation" or "experiment."

In this way there is no means of privileging one text or form of expression over another, no way of distinguishing fact from fiction. Science is just another story. Indeed, the idea that there are facts that can be known, and that some methods are better able to disclose them, is but a means of privileging one practice or method of storytelling over another. It is a means of gathering and asserting authority and power through the control over meaning (which is where this strand meets the Foucauldian one). One gains control by making (or perpetuating already existing) arbitrary linguistic distinctions (ones not based on "fact," for that cannot really be determined) among and within types of texts, and by privileging one set over another through the claim that the distinctions represent reality. This control can occur at the very level of words themselves. The meaning of a word depends on how it is defined in terms of other words. A word differentiates itself from other words in a field of differences. Thus distinctions among words make it appear as though there are real distinctions out there beyond words, and these distinctions among words, and the values that these distinctions possess, can affect what we think the world is like. Those who make the distinctions and assign the values and persuade others that they are important are the ones who possess real power (and power in any form it seems is something to subvert). They can manipulate our views of "reality."

Power pervades language more insidiously. Language often makes distinctions by creating implicit dichotomies, where one side is positive or privileged, and what is left out is then by implication negative, secondary, and even antithetical. The negative side becomes the "other" of the positive. Reality, for example, is something that is often privileged, and those who talk about it or claim to know it are empowered through control over the word. But reality is posited (and thereby "exists") only in relation to the supposition that its "other" exists— which is unreality or fantasy. Similarly, reason, which is taken to be positive and worthy, "exists" only in relation to its other—unreason or emotion or irrationality. Those who can claim to know reality and to use reason are then more powerful than those who do not. And of course the same would be true for moral claims and moral systems. The moral makes itself important by positing its opposite or "other"— immorality or evil. Reason, reality, and morality are simply examples of creating distinctions and prioritizing one set over another that, though still linguistic, create and perpetuate power relations. Meaning and power are mutually dependent. Who controls the "signifier" is all-important, and since there is no truth, no reality to direct us and anchor our meanings, power itself will also be rootless and under suspicion. Moreover, developing techniques—what are called deconstruction techniques—that point out that language is inconsistent, fluid, and multivalent (and that this is equally true for the big and taken-for-granted languages and terms that frame ideas about the real, the objective, and the moral) provides a tactic for destabilizing the dominant power relations.

Puzzles arise as soon as we ask about the motivation behind making these postmodern claims. Here again, I would like to consider the possibility that the motivation for deconstruction could stem from the hope that it can liberate and emancipate us and make us morally better. Before I take up this emancipatory possibility for both of these strands of postmodernism, I must say that there are at least two other and perhaps more obvious motivations. The first is that postmodernists simply want power. Here the moral cynicism that follows from their claims about situatedness and relativism can be turned against them. By posing the virtually totalizing arguments that knowledge is situated and becomes a form of power and that virtually everything is a language game, these very claims then become the postmodernists' way of saying they have a special kind of knowledge; it is a way for them to privilege themselves and their positions. So postmodernism is a means for postmodernists to gain power.

The second position simply takes up the playfulness of postmodernism and sees this to be its object. Nothing that is said, whether by others or postmodernists, should or could be taken seriously, because nothing can withstand the force of postmodern deconstruction and critique. Our intellectual energies then have nothing else to focus on than the inconsistencies of claims, and so our purpose is to enjoy disclosing them, toy with them, and to do so even for postmodernism. Not only is this playful, it also encourages irony, for at the same moment that we hold convictions we see they are inconsistent; at the same time we believe something to be the case, we see this to be without foundation. Hence, the ultimate irony: we cannot live without convictions and beliefs that are themselves groundless. Power is playful. Whatever is posed can be deposed. And since nothing is really more real than something else, the stakes cannot be very high.

These conclusions are plausible and consistent with many of the claims of postmodernism and have been made before.[14] I though would rather consider the possibility that moral relativism, though the likely outcome of postmodernism, is not satisfactory to some postmodernists. They are hoping for something else even though their own framework would not allow them to put it this way. What they may really want without being able to say so explicitly, is to clarify what is really happening, (which means (ironically) they are assuming a realist position) because disclosing the truth will make things better. Here I am arguing that they have an intimation that not only is there a reality (not matter how bound up it is with language), but there is also a good that is not itself a product of the local, situated conditions. I say this because postmodernists often write about injustices, and against power. And many see their efforts at challenging, destabilizing, and deconstructing conventional, taken-for-granted assumptions about meaning and power to be emancipatory. Again, they cannot say how. Foucault avoids constructing a moral theory, and so does Derrida. Yet both have used moral terms and categories in their own works, and several scholars have tried to clarify what might be the moral qualities of the postmodernist point.[15] But again their efforts are virtually nipped in the bud as they try to adhere to its principles that deny such things can be stated positively, or that the good can exist at all. Still, I would like to make a brief attempt to "reconstruct" the moral intimation that I think is behind their efforts. This again is not how a postmodernist would or could put it for the reasons I have already stated, and also for the reason that in my reconstruction, I stress the role of place, which they would not.

The inextricable connection between place and power is the starting point. Place is a manifestation of human power, and it enables us to exercise power. So, if the postmodernist claim is that all of the categories and relations of power are flawed, not only in the sense of not being consistent or even real, but in the moral sense of leading to lies and injustices, then they can contest, destabilize, deconstruct, and even demolish these by contesting, destabilizing, deconstructing, and demolishing place.

In the most general sense, to have a place is already to have a set of rules that define and constrain us, that distribute meaning and power. Place "emplaces." And, here is the underlying moral hope—by peeling away the layers of place, by diminishing its powers, and finally by removing the last vestige of place—the last place—we become truly ourselves. We become somehow better. I am not sure why this would be the case (for as we will see, emancipation does not automatically mean doing good), but I do think this is the key to what may be an underlying moral belief of postmodernism. Perhaps it is that by removing place we can see each other without the distortions of power, of categories, of emplacement.[16] We are then going back to a stage in the geographic state-of-nature thought experiment, to a world where there is only space. There we encounter one another as true individuals, and this is a good thing because then it can allow our basic human goodness to shine through.[17] We each will care for the other, not as a category arising from our emplacement, but rather from our essential and naked humanness in space. Without the efforts of such postmodern deconstruction, everything remains situated, and since situatedness is the source of our problems, then destroying situatedness will make things better. Goodness will rise like a phoenix from the rubble of deconstruction. This at least is how I understand the postmodern motivation, and even the postmodern suggestions that their philosophy and methods can be seen as ultimately progressive and emancipatory.[18]

But what do we make of this? Two things are important to mention here. First, by now it is clear that this emancipation or goodness is not a cultural or social construct. It is not situated, for it is in fact situatedness that has been destroyed with the demolition of place. This means that that we now must assume that there is something that is good and that is also real. This is the only way that removing constraints could actually emancipate us morally. But the existence of the good is not all that is being assumed. It seems also to assume that we, as emancipated humans, are intrinsically good. That is, we now will not simply be free of constraints, but will automatically and invariably choose the good or act out of goodness. Otherwise, once we are emancipated we could do evil as might a sexual predator who is no longer constrained by authority and feels free to follow his predatory desires, thus accomplishing evil rather than good. The point is that emancipation in terms of removing obstacles is good only if those obstacles were themselves impediments to the good, and only if we ourselves are inherently good and/or have intimations of the good and follow it.

Postmodernism of course does not say this, but I believe it can be seen as implied in these strands. The moral theory presented in this book attempts to develop what postmodernism cannot. In doing so it points out how emancipation can be good only if the good exists and if we choose not to deceive ourselves. The need for these two assumptions, and especially the one about choice, is clear when we consider what happens without them. If we believe that after we remove all constraints we will automatically be good because we are inherently good, and it is only the evil qualities of our surroundings that prevent us from being good, we then have the peculiar argument that being good does not require us to decide or choose, and that evil is a product of the environment that we have helped create. But how can we have helped create an evil environment if we ourselves are inherently good? We need not only to have the assumption about choice, but also about the existence of the good, for how can we know that all power and all places are corrupting if there is no understanding of the good as a standard? Wouldn't it, in fact, be more likely that some powers and places are and some are not? Again, to know this, we must have a moral theory that posits what is good. Yet postmodernism cannot articulate this, and so its picture of what would happen is something like a world of disembodied—and nonemplaced—souls. This may be fine in heaven, but as long as we are on earth we will be place-makers, and therefore situated—though not completely. We have the capacity to become less so, and this happens not by destroying all of our situations or places and returning to the geographic "state-of-nature," but rather by using our capacities of free will when these are open to the attractions of the real and the good to create better places.

This ability to conceptually remove ourselves from our situated places, rather than destroying all or virtually all places, is the path to take. I am obviously not saying that places do not constrain and oppress. Rather I am saying that though places have bad (and good) effects, we are never completely their victims. We can and must be able to think ourselves out of them to know if they are bad and what would be better.

MARXISM

The moral import of postmodernism is that while it claims to be about the situated, it actually yearns to demolish it and set us free. The situatedness here has often meant the individual or unique conditions that mold us. Marxism too is concerned with the situated, but here it is the situatedness of class relations that can be repeated over again in many places. The tension between situated and emancipated is dealt with more directly in Marxism. And while one strand of Marx explicitly champions emancipation and its goodness—indeed his project is to help us achieve emancipation—the positive qualities of emancipation and their justification is far less clear and explicit than is his development of arguments explaining how we are situated and how these situated forces construct moral judgments and moral theory—to the point where many critics read Marx and his statements about morality to advocate relativism. Rather than take a side on this issue, I want to consider how his strong case for the social production and the relativity of moral positions overwhelms his development of the goodness of emancipation. In fact, it often suggests that emancipation will come about as a result of these social forces—emancipation will be produced. This is because he is interested in developing a social rather than a moral theory. Choosing the social over the moral weakens the persuasiveness of his vision: he does not take free will and the good seriously enough. It prevents him from more fully articulating his moral view and leads him to expect, in a way that is similar to the postmodernists, that the disclosure of our situatedness will somehow win us over to the moral attractiveness of emancipation.

Marx postulates a vision of what ought to be that appeals to our moral sensibilities. Yet Marx is not offering a moral argument; rather he is saying that these good qualities will likely emerge even if we are not convinced and do not see its appeal. They will do so because of the historical development of political/economic forces. The "ought" will ultimately be a result of what "is." This idea of the unfolding (which can be thought of as the dialectical and materialistic logic of history) makes the condition of free will and our role as moral agents more than a bit uncertain, and leaves the door wide open to the claim that his vision of the "ought" is relative to the unfolding of these empirical conditions.

This vision is found in Marx's description of emancipation and the communist utopia. Marx sees emancipation as a release from the alienating, unjust, restrictive, and oppressive social relations that are found in any class-structured society and in the capitalist mode of production in particular, whose artificial divisions of labor and definitions of work create alienating occupations. Removing these would free us (provide an emancipated freedom) and allow for the individual's self-determination. In communist society the individual will be able to create himself in all of his or her potentialities and totality—to create in unforeseen ways.[19] Human nature lies in this creative potential that would be freed, once these repressive relations are removed.

This creative potential also contains a strong dose of altruism. Producing simultaneously for oneself and for others would release this altruistic component of our essential human nature. "Let us suppose," Marx argues, "that we had carried out production as human beings. Each of us would have in two ways affirmed himself and the other person. (1) In my production I would have objectified my individuality, its specific character, and therefore enjoyed not only individual manifestations of my life during the activity, but also when looking at the object I would have the individual pleasure of knowing my personality to be objective, visible to the senses and hence a power beyond doubt. (2) In our enjoyment or use of my product I would have the direct enjoyment both of objectified man's essential nature, and of having thus created an object corresponding to another essential human nature . . . our products would be so many mirrors in which we see reflected our essential nature."[20] This idea of production for self-fulfillment and for the pleasure of others is echoed in the famous formula: "from each according to his ability, to each according to his need."

Marx assumes that communism and its provision of self-fulfillment and altruism would occur only after capitalism has developed sufficiently to provide this future society with the sufficiently high levels of material wealth so that much, if not all, of an individual's labor would not involve the drudgery of making a living. Also, Marx's ideas about the contributions within communism suggest that there will be no contradiction between an individual's self fulfillment and the common good. These attributes of self-fulfillment and altruism that come with emancipatory freedom are what Marx thought made communism a good system. But he did not spend very much time articulating them. If he had, he would be offering a moral theory. But on the contrary, the social theory weighs in heavily so that it seems these conditions will come about in any case once the unjust economic practices are removed. This means his primary interest is in disclosing the forces preventing emancipation—his primary stress is on the situated.

For him, the economic relations are the principal forces that shape our behavior. They affect other social relations and even our meanings and perspectives. The central social relationship in capitalism that affects everything else is found in the way capital exploits and extracts surplus value from labor. It is in this power of the economic (or the mode of production), and in the relationship between labor and capital, that we find Marx's reliance on the situated. How much of our lives and viewpoints are caused or determined by these economic conditions, and how much free will we have, has been an issue of considerable debate among interpreters of Marx. One can find support in Marx's writing to the effect that we have virtually no capacity to resist economic conditions, or one can find support that we have a greater degree. But the entire logic of Marx's system, its appeal as a scientific theory (though not as a moral program), remains more in tact the less power free will is granted.

Supporting a degree of free will are the specific claims that we can make choices, that we can "make history" (though not under the conditions of our own choosing), and perhaps the most important, the implied argument that we can do otherwise because Marx's writings are not only about explaining what is taking place, but attempting to convert individuals to the Marxist point of view so that change can be effectuated through the will and political action of these converts. But again, countering this on the situated side are Marx's most distinctive and weighty arguments that forces of class conflict (the laws of capital) virtually determine behavior and outlook, that these forces set in motion certain tensions and contradictions that propel us along an historical path seemingly independent of our will, that we can either swim with this historical current or against it, but not in the long run affect it. My point is not to provide the "correct" interpretation, but rather to illustrate how this problem of situatedness and emancipation compare to those in postmodernism and also to the moral theory developed here.

In both postmodernism and Marxism, the situatedness cannot be total, for not only would there be no means of extracting oneself to write about it, or even any point in doing so, but that it is contrary to the way things really are. We do have free will. As we have seen, the post-modernist seems to promote emancipation, but leaves choice out at this juncture, for once emancipated we will all automatically be good because we are innately so. And postmodernism sees practically all forms of situatedness, power, and place as the enemy, and so destabilizing all of them is necessary for emancipation. On this last set of points, Marxism is far more discerning. It takes the view that the most important thing to destabilize and replace is the economic system (which in the most extreme determinism would fall by virtue of historical inevitability), for that is the basic source of injustice; once that is removed, other problems will recede. But what about our goodness? What in Marxism is an emancipated individual and why is that good?

Marx believed that emancipation is of value because in allowing us to do what we want we can become our true, full, authentic, and unalienated selves. Again, why is this good? It is certainly not self-evident. A simple example will illustrate why. Consider this time a rat in a maze. It chooses one direction over another because it is driven to do so by its desire for cheese. Fulfilling this desire makes it a more complete and self-fulfilled rat. The rat is making only small choices. It cannot decide not to pursue cheese, nor can it decide to create other goals. If the rat could really choose in this sense, it would not be a suitable subject for this experiment, for its behavior would be unpredictable like our own. The model of the rat, variants of which form the basis of the most dominant model of human nature, assumes that behavior is caused by concatenations of empirical forces and that choice is simply "selecting" the path of least resistance or most pleasure. This is the model of human beings put forth by most of the social and natural sciences, whether it is sociobiology, modern forms of economic theory, rational choice theory, and psychological theories such as cognitive dissonance and stimulus response.

This idea of choice, as with the rat in a maze, is not the same as free will. If I am the rat in the maze and if you, the experimenter, obstruct my route, I am constrained. And then if you remove that constraint, I am "freed" or "liberated" to pursue my goal, one which I still did not choose in any real sense. Or to change us back into people, if there are no rules or laws or constraints preventing a pedophile—who it is now believed is a person unable to control his actions—from molesting children, he may feel liberated or emancipated, but this is a form of emancipation that most would find morally unacceptable. So liberation or emancipation is not, in itself, a moral category, because the moral still depends on a conception of what it is morally right or good that is not itself a product of circumstances.

The Marxian view goes much farther than the postmodern, but not far enough in articulating what kinds of emancipation are good, why they are good, and how choice must always be involved. And Marx further muddies the waters through arguments about unfolding historical process that will lead us there anyway.

Again, I do not want to argue here with the abstract terms of emancipation, freedom, and justice. My primary point is that to know what kinds of power or what forces need to be weakened so that we can become free of them and morally better already presupposes an intimation of the good. This intimation is being used to select what is oppressive and unjust and from which we should be emancipated, and this same idea of the good would then serve as a guide to our behavior once we are emancipated. Without this moral compass, we may emancipate ourselves from the wrong things, and once emancipated not know what to do. But Marx does not articulate this moral position; his scattered examples of the altruistic and self-fulfilled individual under communism are simply presented to the reader as self-evidently good qualities that seem somehow to arise naturally.

If Marx had developed a moral theory that justified these and other good attributes and that recognized the role of free will in pursuing them, then the plausibility of his moral arguments would themselves have been the principal power that would animate us and not the empirical unfolding of forces. But Marx wanted to replace moral theory and argument with a "science" that identified the underlying material forces creating injustice and propelling us historically. So reason and will became less significant than these material forces that situate us and make us their instruments, and run the risk also of making any moral position relative.

Another difficulty with Marxian theory, which to many individuals is an insurmountable one, lies in the area of validation. As many have pointed out, there are problems with the labor theory of value, with the definition of class, and of course with the role of the individual agent in the process, and there are difficulties with historical predictions: the establishment of socialist states in the 20th century, for example, did not happen in places where capitalism was well-developed as Marx argued it would, but rather in virtually noncapitalist countries. As contemporary proponents of Marxism attempt to adjust the theory to accommodate new forms of the political economy and new social concerns—which include incorporating concerns about fluid and changing ideas of class; allowing a greater role for human agency; suggesting that the realm of meaning (or culture as it is often referred to in the literature) may be virtually as important as, and autonomous of, the forces of social relations; arguing that other types of social relations such as race and gender, and organizations such as Non-Governmental Organizations (NGO's), and multinationals, have great significance in the contemporary process—they may make Marxism more realistic, but at the expense of losing the quality of a theoretical argument, for now there is difficulty knowing what things are more or less important and what drives what. Indeed, it is now difficult to know what a contemporary Marxist is, other than one who believes contemporary society is riddled with injustices and who may (in a very general sense) see the present economic systems as the most important cause. But this is something with which many others might agree. If it is the injustices that are paramount, and the empirical logic of their existence secondary, then we are stressing more the need for a moral theory rather than an economic one to guide us. So, in an important sense, by making Marxism more flexible, contemporary Marxists have gone back to a position that is similar to socialists before Marx—socialists who saw and argued for socialism as a moral force. What is needed then is a moral theory that can make the argument.

Offering a moral theory does not mean ignoring the force of the social. On the contrary, in our moral theory, the focus on place shows that social relations and the moral realm of justice are important, but equally so are meaning and truth, and nature and the natural. The theory however does not tell us which ones are most important overall. Rather the geographical can tell us how they are woven together, and the structure of the loom can provide an inductive means of finding out who is pulling the strings. Most of all, the theory stresses how our capacity to reason allows us to obtain a less situated view which then provides a means of judging our situation and changing it for the better.

What happens in reality when the situated is not tempered by the less situated? What happens when those who want to create a socialist system can focus almost exclusively on the economic because Marxism has not paid sufficient attention to the moral and the role of free will? And what happens when instrumental judgments become the only ones so that "might makes right?" Situatedness and moral relativism can neither judge nor offer resistance to other forms of situatedness that claim to be absolute and totalizing. Indeed relativism frequently shifts into absolutism. And when it does, those involved believe that they are doing good. In our terms, they do not believe they are adhering to an instrumental position at all, but rather have found the intrinsic view. This is the case in the examples that follow. Before we face directly these examples of absolutism, it is important to say something more about how we can know that these ideas of the "good" are mistaken. Of course, a fuller account must await the third section of the book where problems of self-deception are offered as the primary locus for such mistaken views, but our discussion of the circularity within the instrumental logic has itself provided us with the beginnings. It shows that a moral position that coincides with self-interest, or that appears to be self-serving, is a likely candidate. It is by no means conclusive evidence, but it does raise doubts. And if those holding these positions, in the face of challenges, react by making their places less open and accessible through devices such as censorship, so that those within cannot see out and those out could not see in, and thereby further deepen their own situatedness, then it raises even graver doubts about the validity of their moral claims. Let us turn now to examine how forms of absolutism operate geographically.

4

SITUATEDNESS AND ABSOLUTISM

INSTRUMENTAL JUDGMENTS NOT ONLY LEAD TO RELATIVISM, but to the worst kinds of absolutism and tyranny. Their encouragement of a situated position allows the moral strands to be twisted to support a self-serving and self-deceiving view that is absolutistic. Here I will discuss three cases connecting situatedness and absolutism: Nazi Germany (which will be given the most attention), the Soviet Union, and slavery in the antebellum South. I need to say a few words (some of which I have said before) about why these, and how I will address them.

It is important to select cases that have been under moral scrutiny before, because I need to show that a geographical approach to moral theory can shed light that otherwise may have been overlooked, and these three are certainly well-known. They each come with their own so-called "moral theories" that offer "justification" for these systems. Each argues that the virtues of truth, justice, and the natural are situated—and each offers different emphases and interpretations of them: Nazi Germany and slavery in the antebellum South stress the role of nature through the idea of race, and the Soviet Union focuses primarily on the power of social relations (and these differences will be discussed later in the chapter)—but the central point here is that all three come with their own moral justifications. They claim that particular forms of hate, oppression, and violence are inevitable and good.

Hitler for example, in *Mein Kampf*, made it abundantly clear that hate and racism were virtues, and that the true moral perspective was a situated one, drawing on one's own ethnic and geographical rootedness. He claimed that the conventional Western ideas of human rights and its respect for life, liberty, or equality were not essential. As he put it: "There is only one holiest human right, and this right is at the same time the holiest obligation, to wit: to see to it that the blood is preserved pure . . . The *volkish* state . . . must set race in the center of life. It must take care to keep it pure."[1] The Soviet Union provided an ideology that claimed totalitarianism was a virtue, and many in the American South justified the institution of slavery as a good and natural paternalistic system.

But why focus on these rather than the injustices of poverty? The answer is that there do not seem to be any contemporary moral theories or ideologies that argue that poverty and hunger are good, or propose that most of the world should be in poverty and undernourished. Of course, there are social theories that put forth the position that such conditions are the inevitable consequences of specific political economies, as in the social part of Marxist theory that claims capitalism inevitably and necessarily creates these evils, but this is not what capitalist theory claims. On the contrary, capitalist theory, or more precisely theories that draw on classical and neoclassical economics, argue that capitalism will be the tide that lifts everyone's fortunes. On this point, even classical Marxism has admitted that capitalism is a necessary step in the process to socialism, and is an improvement over what preceded it. And in any case, these theories are not about moral choice and intention. Given then the importance of intention and purpose in moral theory, the selection of Nazi Germany, Stalin's Soviet Union, and the institution of slavery makes sense: these examples provide documentation that allows us to explore the question of motivation and intention. Intrinsic judgments, it will be recalled, argue that evil is due to self-deception which results in a lack of awareness. Hence, to take up examples of the evils of absolutism in our later discussion, we need to have some evidence about the psychology of the individuals involved. These examples can lay the foundations.

These are general reasons for the selection of these cases. But I also need to justify why I am going to go into far greater detail about one case—Nazi Germany. To begin with, I want to select a case to illustrate the power of place in greater depth and to serve as a model of how this could be accomplished for the other cases. This means I need to show in some detail, over a range of interconnecting places, scales, and periods of time, how geography is instrumental in the execution of evil. I must provide examples of how places weave and twist elements of truth, justice, and the natural to form a geography that narrows awareness and diminishes variety and complexity. This can take volumes. Here I want to go into enough detail in one case to suggest how it can be done in the others, and still touch on them so that their overall similarities and differences are suggested. To do more would overwhelm a book that focuses attention on a theory of the good.

All three examples provide abundant data, but the Nazi case pushes itself forward because of the evidence it provides on motivations and intentions, which again is central to our moral theory. Nazi Germany not only provides this, but the hierarchical and centralized aspect of the system means this evidence can be directly connected to the creation of the geography of evil itself. We do have some evidence of motivation and intention in the case of the Soviet Union (which too was centralized and hierarchical), but the documents have simply not been as accessible and for as long a time. And though there is much evidence concerning the thoughts and views of slave-holders and others in the American case (as we shall see in Chapter 6), the American system is not a centralized hierarchical one that makes it possible to trace how these thoughts lead to the creation of the system. These then are the reasons I have selected the three examples, and why I devote most of the time to the German case. I offer it as a model for how the analysis can be done in other contexts, including the Soviet Union and the slave system, which I will take up, but only briefly and with a very broad brush.

All three discussions draw on well-known and established facts, but present them according to the way they were assisted by and enacted through place. It is important to say that I am not offering this as the only correct interpretation of these complex systems. Rather I am presenting what I hope is a plausible interpretation that will serve to illustrate how geography undergirds good and evil. And it will be recalled that evil, geographically, involves the opposite of intrinsic judgments: it uses place to narrow our vision and diminish variety and complexity through isolation, tyranny, and chaos. This narrowing and diminishing occurs not only for victims, but also for the perpetrators.

NAZI GERMANY

The Nazi project was racist, nationalist, and totalitarian. This is not only what those who fought against Nazi Germany believed. The Nazis themselves were proud to describe their own position this way. For them, racism and totalitarianism were virtues. They gave each of these a particular meaning, but I want to focus on racism and especially anti-Semitism (which the Nazis defined as a form of racism), for this was the most distinctive quality of the project from a moral point of view. A strong case can be made that anti-Semitism made the Nazi movement different from other forms of Fascism. The Nazi search for a quintessential Germanness in the idea of an Aryan German *volk*, or people, led the movement to define such qualities in terms of what they claimed to be its opposite—the quintessence of which was Jewishness. A deep-seated animosity toward Jews was a long-standing and acceptable part of German tradition, and of other European cultures as well. George Mosse is reported to have remarked that if we were living in the 1920s and were to be told that a European country was going to annihilate millions of Jews, he would have selected France, not Germany, as the likely perpetrator. We though are focusing on Nazi Germany, because it happened under that regime. Concentrating on anti-Semitism should not diminish the fact that other ethnic groups and nations were targeted by Nazism, and suffered terribly under its scourge. Indeed, all "non-Aryan" groups were tyrannized, and Slavs and Gypsies, especially so (blacks were spared for the most part because the Nazis had very little contact with them). Still, the Nazis saw Jews as their principal problem and singled them out for total elimination. (Needless to say, the Nazis provide an example of the essentialist thinking in its extreme.)

Hitler was obsessed with the Jewish "problem" and made anti-Semitism the cornerstone of the Nazi platform.[2] Described in *Mein Kampf* as subhuman and in the utterly negative terms of a virus, disease, and a pestilence, the Jew was nonetheless a formidable opponent. "The mightiest counterpart to the Aryan is represented by the Jew," says Hitler. The "Jewish menace" is the most serious threat facing Germany, sapping this people's strength as would a "vampire" or "parasite."[3] Other Nazis echoed this view. Heinrich Himmler, Hitler's Minister of Interior, argued that the Jews, who he claimed were "spiritually and mentally much lower than any animal," must be exterminated, and in a manner that would not harm "the soul of [the German] people . . ."[4] and Reichsmarshall Göring, said "This is not the second World War; this is the Great Racial War. The meaning of this war, and the reason we are fighting out there, is to decide whether the German and Aryan will prevail or if the Jew will rule the world."[5]

Anti-Semitism was the most important thing on Hitler's mind even as he faced defeat and suicide. His last recorded statement dictated in his Berlin bunker on April 28th and 29th, 1945, the last two days of his life, when everything before him was crumbling, reiterated his hate for the Jews and his conviction that the entire war was about destroying this enemy: "Centuries may lapse, but from the ruins of our cities and monuments will rise anew the hatred for that people to whom we owe all this, they who are ultimately responsible: international Jewry and its acolytes. . . . Above all, I obligate the leaders of the nation and their following to a strict observance of the racial laws, and to a merciless resistance to the poisoners of all peoples, international Jewry."[6] The mission of the Nazi party was to rid first Germany and then the world of Jews. For Hitler, World War II, was in no small measure a racial war, and the Jews were the primary enemy.

How do we describe this process of elimination in geographical terms? This will require three steps. First, I will describe the roots and historical/geographical transformations of anti-Semitism as a cumulative process, beginning as an issue about religious truth, and then adding issues about justice, and then the natural. These categories of truth, justice, and the natural apply terminologically, but only in that sense, for anti-Semitism as a doctrine of hate so twists the meanings of these terms that they now become their opposites. What is called truth is in fact a lie, what is claimed to be just is unjust, and that which is called natural has no foundation in nature. But the anti-Semites are not only blind to this, but accuse the Jews of being a lying, unjust, and unnatural group. To the Nazis, the Jews embody these "disvirtuous" threads, while the Nazis embody the real and virtuous ones.

The second step expands on the recent, and especially the Nazi, addition of the thread of nature to the weave and what this meant. The central issue here was that German identity and the German volk were associated with a natural rootedness in German place and soil. Having a place, and especially the German place, produced the highest virtue. Jews, on the other hand, were uprooted and placeless which diminished their capacity to be human. As the Nazis rose to power, they fully intended to create a perfect German landscape, with an appearance or surface that was both aesthetic and devoid of Jews. The effort to create this surface led to horrifying rules and flows that were the geographical basis of the Holocaust. This proceeded along two paths: expulsion and spatial concentration.

The third step focuses on the crucial aspects of concentration, its ever-tightening grip of rules, constricted flows, and dwindling variety that culminated in the autarkic, transgressive, and tyrannical landscapes of evil of the concentration camps. After discussing these three issues, I will turn then to the function performed by everyday places such as the home, the school, and the church in perpetuating the Nazi ideology; and then conclude with a discussion of sites of resistance and goodness.

Three (dis)Virtues

Anti-Semitism was not of course a Nazi invention and did not exist in a vacuum. It evolved over nineteen centuries during which it moved from questions of truth, to truth and justice, to truth, justice, and the natural or racial. How then did the first issue—truth—emerge?

Truth   Anti-Semitism is fundamentally a product of how many Christian leaders viewed the relationship between Christianity and Judaism. Jews present Christians with a serious intellectual problem. Christ and most of the disciples were Jews. In the Old Testament are predictions that a Jewish messiah (prophet or leader) will come to save the world. Given that the Christians are convinced not only that the Messiah arrived in the form of Christ, but that Christ is also the Son of God, they are deeply troubled that the Jews, whom they recognize as the original authorities on this subject, do not share this belief.

Since their religion stems from Judaism the Jewish view cannot be dismissed. This leaves the following alternatives for Christians: the unacceptable one that the messiah that God promised has not arrived; or the more palatable alternative that the Jews are wrong. Clearly, Christians accepted the second one. But putting the alternative this way does not convey its moral significance. This comes about when Christians ask how can the Jews be wrong? For if the Christians believe that Christ is God and the Jews do not, then the Jews are blinded to this truth. Have the Jews been deceived or enticed by the devil not to believe? Why else would they not see the truth and convert? What then should good Christians do? Seeking converts has always been an important strategy of Christianity, but it was particularly pressing that Jews be converted, for it not only would save them from sin but also convince the Christians of their own righteousness. If they could not be converted then they must be discredited through attributing any and all problems to the Jews. And so they became the classic scapegoat.

Justice   Anti-Semitism is not about the characteristics of Jews but about the need for a scapegoat. Instead of focusing on why anti-Semitism is a grave injustice, I want to understand how those who were anti-Semitic did not see it that way, how they believed that their position was right and how, in the process, issues of social relations were added to concerns about truth.

Since the Jewish diaspora began, with the destruction of Jerusalem in A.D. 70, the Jews were an uprooted people. Their spiritual life and religious beliefs required a complex set of social customs. As Christianity became more distinct from Judaism, the two diverged socially as well as religiously. Jews and Christians tended to live apart. Jews held their own holidays, followed a complex dietary code, worshiped in Hebrew, practiced circumcision, followed different laws about marriage, divorce, inheritance, and usury, often meted out different punishments for crimes, and in an age where people wore clothes that were emblems of their position in society, Jews dressed differently. These differences in social relations and conceptions of justice were added to religious animosities and made it increasingly dangerous for Jews to live in Christian communities. By the late Middle Ages residential segregation in ghettos became imposed by Christians who did not want contact with Jews and by Jews who wanted protection from Christians. (Ghettos ended in Europe in the 19th century, but were revived by the Nazis.)

Even in ghettos, Jews were not part of place in a permanent sense—they were not rooted—for ghettos existed only at the forbearance of Christian authority. Jews could not live outside, and so they were not allowed to be part of the landed classes. In a time when the predominant form of livelihood was agricultural, Jews were forced to earn their livelihood in crafts, trade, and commerce. A primarily rural, provincial, feudal, Christian Europe would naturally look askance at these practices; and when coupled with their lack of any "homeland," the Jew became even stranger and more foreign than the general urban and cosmopolitan world of traders. These social stigmas became as important to anti-Semitism as the religious distinctions between Christian and Jews.

Religious and social differences were threads running through all of anti-Semitism. But in the 18th and early 19th centuries, more complex social issues about Jews developed with the rise of nationalism and were given particular twists in 19th century Germany.[7] Using allegiance to a territorial state to form one's national character and identity was a common geographical practice in the 18th and 19th centuries. People came to think of themselves not so much as a resident of this or that village or town, or as a member of this or that trade, but as an Englishman, a Frenchmen, or Spaniard. This was also becoming the case in Germany by the 19th century, but the Germans added other and distinct components to their social definitions. An extremely important one was the idea of the volk. Volk was even more powerful than nation in providing a feeling of warmth and identity in a modern world that offered the cold abstraction of justice and humanity, because volk could condense into one people, landscape, and place the all-powerful and universal forces of nature. Volk channeled the vastness of nature into the topography and soil of an area, and thus into the body of the individual who resided there, who ate the food grown on the land, and so had the soil course through his or her blood. The condensation of nature into place, and nature's infusion into the bodies and lives of the groups inhabiting it, allowed the individual to become part of a local group that possessed cosmic significance. Different volk would possess different characteristics in different places. In the German case, the *volk* expanded into something like the German territorial state, and this allowed the feelings of identity and belonging that supposedly come about as part of a clan, tribe, or local community to now be transferred within the larger geographical scale of the nation. This "natural" and "harmonious" link between people and place forged the spirit and soul of the individuals who comprised the volk. It gave them a shared national character that was different from other volk and nations. It was claimed by its proponents to be a resurrection of an ancient connection and was constantly referred to by the term "rootedness." Those who were uprooted or dislocated must therefore have impoverished souls and could not be part of this group. The volk provided a rationale for keeping the society closed, and of even discriminating against those in the city who were more mobile and uprooted from the land.[8] In a peculiarly circular way, it also gave the Germans a justification to expand their territory: if ancestors of current Germans lived somewhere else, as in present Slavic areas, that then proves that these places too must be providing qualities that sustain the volk, and this then justifies the recovery of that place.[9] As one Nazi official was to say, "the idea of blood and soil gives us the moral right to take back as much eastern land as is necessary to achieve harmony between the body of our people and geopolitical space."[10]

To be German then, was not only to be non-Jewish, and non-French (or Italian, or Polish) but also to be part of the soil, the landscape, the place of Germany, and thus to be a member of its volk. People and place became intimately connected. Even though other social groups and volk were inferior to the German or Aryan, they at least had their place. Jews, on the other hand, were supposedly completely without roots. This made them "unnatural."

The Natural The concept of the volk and its relationship to land, soil, and blood already intermixes part of what the Germans took to be the natural into the realm of the social. From the German point of view the relationship is not intermixing, but reducing. They wanted science to explain and justify their anti-Semitism. (Most Nazis then were able to transfer the emotion of hate to the cooler realm of science and reason. Admitting one was anti-Semitic was simply saying they were scientific. This explains how a person like Höss—the commandant of Auschwitz—was able to decry anti-Semitism, but only the emotional kind, which he saw as obscuring and doing damage to the infallible reasoning behind scientific anti-Semitism. In his memoirs,

after the Nüremberg trials, he claims: "Today I realize that the extermination of the Jews was wrong, absolutely wrong. It was exactly because
of this mass extermination that Germany earned itself the hatred of
the entire world. The cause of anti-Semitism was not served by this act
at all, in fact, just the opposite. The Jews have come much closer to
their final goal.")[11] Science was gaining enormous prestige in the modern period, often at the expense of religion, and an attraction of the
volk concept was that its non-religious dogma could appear to be
grounded in "scientific" principles. The volk concept and its connection to place were often framed in Darwinian and Social Darwinian
terms. Culture, like biology, operated along the lines of a survival of
the fittest. A volk was like a species that survived and thrived in a particular habitat, and some volk or species preyed on and displaced
others. The volk then were a "race" that biologically transmitted its
cultural characteristics; the Jews, on the other hand, were a "race" that
could not develop a truly human culture because they had no place.
The more one linked volk to race and biology, the more anti-Semitism
gained currency. It was no longer a matter of meaning, or a matter of
social relations and justice. Now it was a matter of nature and fact. It
was scientific. The volk were not only a cultural, linguistic, historical,
and geographical entity, they were also a biological unit—a race that
could be further genetically purified and so fulfill its biological destiny,
or sullied by breeding with inferior races. Of course, a reduction of
concepts to "science" and biology could never be complete even in the
fantasy world of racist biology. The religious and social genesis of anti-
Semitism still played as important a role as ever. Even so, the way of
expressing anti-Semitism relied more and more on biological analo-
gies and metaphors.

Jews were being described in biological terms as a disease or a parasite on the body of the volk that sapped its strength and poisoned its
blood. In *Mein Kampf*, Hitler asks: "Was there any form of filth or
profligacy, particularly in cultural life, without at least one Jew involved in it? If you cut even cautiously into such an abscess, you
found, like a maggot in a rotting body, often dazzled by the sudden
light—a kike. . . . This was the pestilence, [albeit a] spiritual pestilence,
worse than the Black Death of olden times, and the people was being
infected with it."[12] Because Jews were likened to disease, "a parasite in
the body of other nations and states . . ."[13] persuading Jews to convert
was the not the solution, for that would be mistaking a difference in
meaning for the fundamental problem—the natural or biological.
Indeed, Hitler criticized most vituperatively those who thought that
reason and conversion were the best way to combat the so-called

Jewish threat.[14] Racial mixing, rather than disobedience to God, became the new original sin: "Blood sin and desecration of the race [or racial mixing] are the original sin in this world and the end of a humanity which surrenders to it."[15] These biological disorders of the Jew were the underlying cause of their social and intellectual impoverishment. Race is the reason "the Jew possesses no culture-creating force of any sort."[16] Race impoverishes not only their social relations and sense of justice but their ideas as well. These degenerate ideas, according to Hitler, include espousing democratic principles, a cosmopolitan outlook, and a commitment to the dignity of all human beings. These are mistaken and degenerate because they deny the real conditions of difference and competition among groups that stem from the biological basis of race. Universal rights are obscene: there is only one real human right and this is to preserve racial and volkish purity. "The highest purpose of a volkish state is concern for the preservation of those original racial elements which bestow culture and create the beauty and dignity of a higher mankind."[17] These are Hitler's words, and not surprisingly exactly the same conception of the Jew is found throughout the Nazi party.

These volkish and sociobiological arguments in German anti-Semitism should not make us lose sight of the fact that as early as the second half of the 19th century, Germany began granting Jews political and economic liberties. This came about as part of the power of the enlightenment and its particular doctrines of inalienable human rights that were drawn from the American Declaration of Independence and the principles of the French Revolution. These were the values espoused by most of the more educated in the population, and it was partly in reaction to the strength of these views that the volkish and Social Darwinian theories developed. So it ought not be surprising that even though racism was popular, many in control of the government and influenced by the Enlightenment philosophy, (and, one should add, simply inspired by real moral impulses) were enacting legislation that increased the freedom for the German Jews.[18] In 1807, Jews were "emancipated" in the first German state and were extended total civil equality during 1869–1871 throughout Germany.[19] But even the liberal supporters tended to present the Jewish case for emancipation and citizenship as a means of bringing Jews into the mainstream by making them more German and less Jewish. If German society stopped discriminating, eliminating the oppressive environment that surrounded the Jews, they would respond by no longer having to manifest their negative qualities, which after all were simply a means of defending themselves.[20] The liberal position, as Goldhagen put it, was "we will defend you, so long as you stop being yourselves."[21] But most importantly, the legal freedoms that were granted and in place in the early 20th century did not reduce the anti-Semitic quality of German ideology. When Germany experienced difficulties, and scapegoats were sought, the Jews were the victims, but this time the successes of their assimilation were added to the list of grievances, and, with a new twist that went like this: Germany had not only accepted Jews, but in doing so had diluted volkish philosophy with the insidious abstract universal values of Judaism, and this resulted in a weakened volkish resolve that could not muster the will to resist Jews as citizens and keep the volk pure. Though the Nazis stressed the "natural" side to anti-Semitism, they still employed other realms to cover all the bases. This gave their arguments the quality of a moving target. If one objected to the anti-Semitic argument that was grounded in one of these realms (such as Jews are evil), the anti-Semite could easily shift to another.

Expulsion and Concentration

These then were the elements or (dis)virtues that were woven by European, then German, and finally Nazi anti-Semitism, in numerous places and at varying scales. These also were the threads that were thought to constitute the Jew-as-a-self. This weave animated further place construction as the Nazis manipulated rules, flows, and surfaces or appearances of places to isolate and destroy the Jewish contaminant, and create a pure Aryan landscape. We turn now to the Nazi efforts to use place for these ends. Especial attention will be given to the rules of in/out for these are far easier to document in laws and legislation than are the spatial flows and the contestations over appearance and reality.

The Loom and In/Out of Place Rules In the early 1930s, on the heels of the humiliating Versailles Treaty, a worldwide depression, and the ineffective Weimar Regime, smoldering German anti-Semitism was ready to ignite and provide Germany a scapegoat, and Hitler. The Jews, Hitler declared, were behind all of Germany's problems. The defeat of Germany in World War I was due in large measure to the alleged fact that virtually all of German "production was under the control of Jewish finance. . . . The spider was slowly beginning to suck the blood of the people's pores."[22] Hitler's "war against the Jews" as Lucy Dawidowicz calls it, began in earnest on February 24, 1920, when "just six months after the Weimar Constitution was enacted, the NSDAP issued its twenty-five point program, which asserted that no Jew could ever be a member of the German Volk, that only persons of German blood could be regarded as members of the Volk and citizens of the State."[23] There is no doubt that anti-Semitism was the heart of the national socialist platform.

Conversion to Christianity, or its modern form of assimilation to German culture, was no longer an option. The Nazis focused on expulsion and annihilation. When exactly annihilation became *the* Nazi objective is difficult to say. It became an explicitly articulated (though highly secret) policy by August or September 1941, when Hitler approved the Final Solution, as the annihilation was called, and dedicated most of the concentration camp system to this end, but there is strong evidence that this policy was brewing for many years. As early as 1922, Hitler was asked what he would do to the Jews once he assumed full power. He answered: "Once I really am in power, my first and foremost task will be the annihilation of the Jews. As soon as I have the power to do so, I will have the gallows built in rows—at the Marienplatz in Munich, for example—as many as traffic allows. Then the Jews will be hanged indiscriminately, and they will remain hanging until they stink; they will hang there as long as the principles of hygiene permit. As soon as they have been untied, the next batch will be strung up, and so on down the line, until the last Jew in Munich has been exterminated. Other cities will follow suit, precisely in this fashion, until all Germany has been completely cleansed of Jews."[24]

At this point annihilation may have been more of a fantasy or wish than a plan. A real plan would have to develop slowly for several reasons. The idea was so horrible that it must be kept secret until Hitler was securely in power and had confidence that it could be achieved. He must gain the trust of a vast majority of the citizens of the country, for many Germans would be needed to assist in carrying out the complex process of exterminating so many people, and those Germans not involved must at least not offer resistance. To kill so many, he must have in place the technological apparatus, including a multiple-layered geographical system that could collect, isolate, and then destroy the Jews not only in Germany, but in the rest of Europe. An ironic but important obstacle was the German obsession with following laws. Even those Germans who were strongly anti-Semitic still prided themselves on being law-abiding. Building a system for annihilation would require that it be supported by such a system and that would take some time to build.

Until these factors were in place, the principal approach would be to geographically isolate the Jews and prevent them from mixing with Germans. This was accomplished by the dual strategies of geographic isolation and expulsion. Their differences were really a matter of geographical scale. Expulsion "cleansed" the country by forcing Jews out, and isolation "cleansed" (most of) it by concentrating them within. During the later 1930s, after the war began and Germany occupied areas of Europe with especially high Jewish concentrations, expulsion in the usual sense was no longer an option (for where were they to go?): they could be collected and then perhaps expelled to one gigantic camp like Madagascar, where they would be caged in and left to die; or they could be collected and eradicated within Europe. The Final Solution involved the latter.

A technical obstacle was to define what it meant to be German or Aryan and Jew or non-Aryan. Shortly after the Nüremberg Party Rally of 1935, a classification of degree of Jewishness was established and remained in force until the collapse of the Third Reich. This had the effect of creating three classes—those "descended from two Jewish grandparents belonging to the Jewish religion or married to a Jewish person on [or after] September 15, 1935, and persons descended from three or four Jewish grandparents; those descended from two Jewish grandparents but not belonging to the Jewish religion and not married to a Jewish person on [or after] September 15, 1935; and those descended from one Jewish grandparent. Initially, the laws affected the first group most severely, but as the power of the regime was consolidated, the second and third were treated more and more like the first."[25]

Once it was clear who they were, the next and most crucial step was to purify Aryan society and German land by ridding it of Jews. The resulting "purified" surface would also be an "aesthetic" one. Part of this depended on the appearance of its inhabitants, and Aryans were supposed to be tall, blond, blue-eyed, and handsome. The landscape was to be populated by them and was to be predominantly rural because city life, even for a German, could be uprooting—as the Nazi chief ideologist, Rosenberg, wrote in his *The Myth of the 20th Century*, "Today we see the steady stream from the countryside to the city, deadly for Volk. The cities swell ever larger, unnerving the Volk . . . and destroying the threads which bind humanity to nature; they attract adventurers and profiteers of all colors, thereby fostering racial chaos."[26] This agrarian romanticism required the planned arrangement of landscape to recreate and perpetuate the close connection between German blood and soil. Even newly conquered lands in the East, according to Himmler, could be made German by creating such an appearance. "If, therefore, the new . . . living spaces are to become a homeland for our settlers, the planned arrangement of the landscape to keep it close to nature is a decisive prerequisite. It is one of the bases for fortifying the

German Volk."[27] This would involve German-looking farmsteads and hamlets that, Himmler believed, "should contain fields separated by groves, trees, hedges, and shrubs [for these were not only neat and tidy, but also were expected to] provide a habitat for weasels and hedgehogs, buzzards, and falcons—the farmer's allies in his war against mice and vermin. Himmler also speculated that such changes in the landscape would create protection from the wind, increase dew, and stimulate the formation of clouds, force rain, and thus push a more economically favorable climate farther towards the East."[28]

Cleansing the landscape and creating a unified and pure environment—a real Aryan surface—meant removing the Jews, first by expulsion and concentration, and then by annihilation. This required carefully crafting a complex web of rules of in/out of place that would alter the spatial relations and interactions of Jews and Germans at all geographical scales. The new rules would either push Jews out (emigrate) or concentrate them. Having them leave Germany required that Jews themselves see how dire the Nazi threat was and that other places accept them. Both of these were difficult. Most German Jews thought of themselves as Germans. Families had been completely acculturated and Jews had resided in German territories for nearly a thousand years. Also Germany was among the most enlightened and advanced societies of its time. So they asked: How could this really happen? Surely it would pass: Hitler would either change, or the Germans would come to their senses and throw him out. This was what many Jews thought. And those who were convinced that they should leave had difficulty finding countries to accept them. After all, anti-Semitism was not only a German problem. Still, many did leave, but by 1938 over two-thirds of German Jews remained in Germany.[29] They had weathered the Nazi storm and simply refused to believe that matters could grow worse. "In addition, it was becoming progressively more difficult to gain entrance to other countries. Western nations were alarmed at the prospect of opening their doors, not only to the Jews of Germany but to the much larger Jewish communities of countries such as Poland and Romania."[30] In a 1936 speech, Chaim Weizmann, who after the war became the first president of the State of Israel, described the problem of a place for Jews this way: "There are . . . [in Europe] six million people doomed to be pent up in places where they are not wanted, and for whom the world is divided into places where they cannot live and places where they cannot enter."[31]

In the meantime, the in/out of place rules progressed methodically to isolate, and continuously shrink the area the Jews could occupy. There were over 2,000 laws enacted to narrow their world and to prevent Aryans from having contact with the Jews—laws preventing Jews from being farmers, public servants, or practicing the professions, and from teaching and attending non-Jewish schools and universities, and laws preventing Germans from patronizing Jewish enterprises. Many of these laws may not have mentioned place, but they were all linked to them. This legal vicelike grip on the spatial interactions, and rules of in and out was only the beginning of the process of isolation. Shortly after the Austrian Anschlüss (March 12, 1939) and the invasion of Czechoslovakia (September 1, 1939), more straightforward means were undertaken to limit the geography of Jews. An important step was the September 1, 1939, act of forced "relocation" of Jews in "communal houses." And on the heels of this began the walling off of Jews in ghettos. In May 1, 1940, Lodz Ghetto, containing 160,000 Jews was sealed off. In October 1940, Warsaw Ghetto walls were built to isolate its 400,000 inhabitants. As ghettoization continued in 1941, large-scale and systematic massacres of Jews began. Hundreds of thousands of Jews were killed simply by rounding them up and shooting them. But this was found to be too inefficient. It was difficult during the process to keep large crowds of victims under control, and disposal of the bodies presented a serious problem. A temporary solution was first employed in June 9, 1942, in Riga, Latvia: the implementation of mobile gas vans for exterminating Jews. The permanent solution was of course the gas chamber in the concentration camp.

Focusing on the enactment of in/out of place rules does not provide details of how the spatial interaction and surface and depth components were engaged. We will await the discussion of the concentration camp to examine all three but even with only references to the laws separating Jews mentioned above, it is possible to imagine how, for example, the enactment against Jews owning farmland not only changed spatial interactions—rural Jews now had to move to the cities—but also reinforced the appearance (the surface/depth component) that Jews were footloose, urban, and out of place, and that they did not belong to the real rural Aryan landscape. The farms left behind would be claimed by Aryans (which would again change spatial interactions) and bring the appearance of the Nazi landscape closer to its ideal; or, we can imagine that the restriction of Jews attending German schools (because of in/out of place rules) not only meant that these Jewish children had to move to different educational sites (spatial interaction), which required that Jewish adults then set aside places for the education of their children and participate as instructors, but that German schools were now racially purified. The "landscape" within the school was cleansed (surface/depth), and the separation of Jews was soon accepted as normal and natural. Their difference was reinforced by their physical separateness, which in turn reinforced Aryan prejudices about reality and truth. Now that Jews could not go to Aryan schools, they could not learn what was real and true (as defined by the Nazis), but only a corrupted (Jewish) version of the truth.

All of these rules were geographic strategies that the Nazis weighed very carefully, thinking through their implications according to an implicit understanding of the structure and dynamics of place. Two brief accounts show just how much time and attention were given to the implications of each rule. In a November 12, 1938, meeting attended by Propaganda Minister Goebbels and Hermann Göring (two of the highest-ranking Nazi officials) and Heydrich (Himmler's right-hand man), details about the concentration of Jews were discussed. One of the issues concerned housing. Goring brought up the question of whether Jews should now be crowded into ghettos: given all of the geographic restrictions enacted thus far, should they now jump to ghettoization or continue these incremental stages of isolation? The ghetto could quickly "clean up the landscape" by providing a large, densely inhabited place, but this had its drawbacks, for concentrating too many Jews together could create the possibility of a rebellion. This problem of resistance worried Heydrich, who argued that "from the point of view of the police, I don't think a ghetto in the form of a completely segregated district where only Jews would live, can be put up. We could not control a ghetto where Jews congregate amid the whole Jewish people." Indeed, having Jews somewhat more dispersed, still living in Jewish houses or blocks, but not in large masses, would allow the Germans greater opportunity for surveillance. Being amidst Germans or at least under their gaze could "force the Jew to behave himself. The control of the Jew through the watchful eye of a whole population is better than having him by the thousands in a district where I cannot properly establish a control over his daily life through uniformed agents."[32]

Even the geography of such small-scale activities as the seating on trains commanded the attention of the Nazi command. At the same meeting, Propaganda Minister Goebbels complained that "it is still possible today for a Jew to share a compartment in a sleeping car with a German. Therefore, we need a decree by the Reich Ministry for Transport stating that separate compartments shall be available for Jews; in cases where compartments are filled up, Jews cannot claim a seat. They will be given a separate compartment only after all Germans have secured seats. They will not mix with Germans. . . ."[33]

Rules of place were not only enforced by the power of the state, they were also at least tacitly supported by the general populace. How complicitous was the German citizenry in this project is difficult to say. Goldhagen believes that most Germans were "Hitler's willing executioners," yet the Nazis went to great pains to keep the most heinous acts, such as the death camps, a secret from the general populace. Here we can safely assume that the Nazi command was relatively confident that most Germans would not object to the vast majority of these in/out of place rules. Indeed, Heydrich's point that the German citizenry could act in effect as prison guards for a more dispersed form of ghettoization suggests that they thought the German population as a whole could be relied upon to enforce Nazi discrimination laws.[34]

Partly for the reasons mentioned by Heydrich, ghettoization was postponed and Jews were concentrated in houses and blocks. But as Germany pushed farther east, and more Jews from occupied areas were now under Nazi control, this tactic was to give way to formal ghettoization, a process that was to be in full swing by 1940; however, even ghettoization was only a step in the process to annihilation. Heydrich, in a 1939 directive, explained how ghettoization was to be an interim measure necessary until the concentration camps were ready to receive the Jews. These concentration camps were part of what he referred to as the "final aim" of the process. "A distinction must be made between the final aim (which will require extended periods of time) and . . . the stages leading to the fulfillment of this final aim. . . . For the time being, the first prerequisite for the final aim is the concentration of the Jews from the countryside into the larger cities. This is to be carried out speedily . . . [and once they are in ghettos, there will be regulations] which will forbid their entry to certain quarters completely and that they may . . . not leave the ghetto, nor leave their homes after a certain hour in the evening, etc. . . ."[35]

The ghetto   The process of ghettoization began in earnest in the occupied areas, with Heydrich's Sept. 21, 1939, order. On December 10, 1939, the first major ghetto, that of Lodz, was planned. Uebelhor, the German in charge, echoing the interim nature of the ghetto said that "the creation of the ghetto is, of course, only a transition measure. I shall determine at what time and with what means the ghetto—and thereby also the city of Lodz—will be cleansed of Jews. In the end, at any rate, we must burn out this bubonic plague."[36] But mechanized and efficient extermination would await the next stage—the camps.

The Germans selected a run-down area of Lodz in which to concentrate the victims. There were already 62,000 in that neighborhood, but soon 100,000 more were moved in from other parts of the city. Individuals from a given neighborhood were allowed a few days to leave their dwellings and move to the ghetto, and after the resettlement had been completed, the Germans enclosed it with a fence. There were eventually to be about 144,000 Jews living in this small 1.6 square mile sector of the city. Shortly after, the infamous Warsaw ghetto was erected with 445,000 inhabitants in 1941 in an even smaller area of 1.3 square miles. It too was cut off from the rest of the world.

By the end of 1941, almost all Jews in the occupied areas were in ghettos, and each ghetto was sealed off. The ghetto, and the starvation, disease, and neglect that it promoted, could have served as the mechanism of extermination (estimates are that 800,000 Jews died in them)[37] but for the Nazis, it was too slow and unreliable. To speed things up, they had special branches, such as the Einsatzgruppen and other "security" units, run experiments in large-scale systematic killings. The Einsatzgruppen, formed at the same time the Germans attacked the Soviet Union, were mobile killing units whose purpose was to cleanse the newly conquered territories by killing as many Jews in these areas as possible. This system of mass killing was new, so the orders were to go slowly and recruit locals to do the actual assassinations, for it was not clear if the Germans could stomach the brutality without having some time to "adjust." This involved rounding up a few hundred to several thousand people and shooting them on the spot. At the very beginning mostly men were selected, but very soon, the troops became inured, and women and children, were included. Even so, these were "small-scale" mass killings compared to what was to be.

Soon the scale was expanded with the addition of SS and army units. The resulting deaths were now in the thousands. All told, approximately 1,300,000 Jews died in these "open air" killings.[38] This, though, was still not good enough. It created problems of speed, of efficient and complete removal of bodies, secrecy and disposal of belongings, and it took its toll on the soldiers, who though seemingly willing, were nevertheless often psychologically damaged by what they did. So a faster, and perhaps most important of all, impersonal alternative was sought.[39] The solution was gassing. This technique was first used in mobile units (first by simply pumping the carbon monoxide exhaust back into the trucks—but it was found that too many were still alive by the time they arrived at the grave sites—and later by using Zyklon B); but permanent sites were preferable because of the logistics of assembling people and disposing of the bodies, and of keeping this mostly out of sight. The solution was the gas chambers in the concentration and death camps.

The camps   The ghettos were the antechambers to the concentration camps, where the final stage of the final solution would occur. The word concentration camps really applies to a wide range of brutalizing places of incarceration that began almost as soon as the Nazis took power, and not all of them, even during the War, were places of extermination. This does not mean that people escaped torture, hunger, and death in any camp, but that only some of them were designed for assembly line killing. It was in these that the extermination took place in earnest (between 3–4 million Jews died in them, from a total of 5–7 million Jews killed by the Nazis, out of approximately 8 million Jews in Europe).[40] These camps fell into two groups. One we can call the *death camp* which attempted to funnel every prisoner directly and immediately from the cattle car train to the gas chambers; and the other we can call the *concentration-death camps* where many were killed as soon as they arrived (as in the death camp), but a few were culled for a slower death through inhuman labor and torture. It is important to remember that Jews were not the only victims in the concentration-death camps, and for the entire system, some estimate that at least 18 million Europeans passed through the system and at least 11 million died in it.[41]

To understand the role of the concentration camps as a system, and in the more particular role of the concentration-death camps and death camps, we should step back for a moment and think of these as the most extreme geographic examples among many instruments of racial and cultural purification devised by the Nazis. Concentration camps were places of incarceration for individuals who were not "criminal" in any conventional sense of the word—they were then not prisons—but rather for acts, conditions, and states of affairs that were claimed to be contrary to *any* goal of the Third Reich. This would include one's political affiliation (namely, anyone opposing Fascism and/or anyone who is a communist), one's ethnic and cultural identity (anyone who is not "racially pure," such as the Slavs), and then of course the Jews. The camp system overall was designed from the very beginning to remove such individuals from the rest of German society. Dachau, founded in March 20, 1933, almost immediately after the Nazis took power, was the first of such camps, and was later to be a concentration-death camp. In a pep talk, the commander addressed his SS underlings who administered the camp, saying that "You all know what the Führer has called upon us to do. We haven't come there to treat those swine inside like human beings. In our eyes, they're not like us; they're something second-class. . . . Any man in our ranks who can't stand the sight of blood doesn't belong here, he should get out. The more of these bastards we shoot, the fewer we'll have to feed."[42]

According to Goldhagen, the camp system was the singular invention of the Nazis, their earliest, largest, and most important institutional innovation. (Yet there is evidence that Hitler was inspired by the Soviet gulag system.) They were a powerful, though often disguised, presence on the Nazi landscape—there may have been 10,000 thousand in all.[43]

From 1933–1939, camps continued to be constructed, but their infamy was to come from the push for the Final Solution with the creation of the concentration-death camps and death camps. It was the new techniques of gassing that made these feasible. "Each of the concentration-death camps and death camps would exterminate primarily the Jews of the surrounding regions. . . ."[44] The prisoners were collected and moved by special trains and cattle cars under conditions that were not unlike those of the camps themselves. And the concentration-death camps were connected to satellite camps that provided slave labor for particular industries in the surrounding areas.

During the War, 13 "official" main concentration-death camps (most of which contained dozens of satellite camps) were operating—Auschwitz/Birkenau, Majdanek, Dachau, Sachsenhausen, Ravensbruck, Buchenwald, Flossenburg, Neuengamme, Gross-Rosen, Natzweiler, Mauthausen, Sutthof, Dora/Nordhausen—and four death camps—Chelmno, Belzec, Sobibor, and Treblinka. Each concentration-death camp possessed similar features. They tended to be rural, in an area accessible by rail, within the region that was to provide the inmates. Removal from direct view of ordinary citizens was important but not essential (Sachsenhausen was only 30 kilometers north of Berlin and within the midst of a settlement and Auschwitz was adjacent to the city of that name), but the four death or killing camps were more remote and secret.

In most cases, the concentration-death campgrounds consisted of two areas. One was for the housing of the guards and the families of the higher-ranking officers. The other was the concentration camp proper. The two were functionally interconnected, as is the case with the geography of workplace and home elsewhere. But here the relationship was particularly revealing. The workplace was a site of horror, and those running it wanted to create the illusion that this was not a part of their real life. Rather their work was something like an illusion. Reality was found in their residential environment. The SS were housed in pleasant and often luxurious surroundings that attempted as much as possible to replicate an image of "normal" German family life.

At Buchenwald the higher-ranking officers were housed along a road, at the end of which were "ten luxurious villas equipped with every comfort . . . these tasteful wooden houses had massive basements, garages of their own and wide terraces with a magnificent view."[45] The lower-ranking officers lived in "handsome one- or two-family houses, each with its own garden . . ."[46] and the troops were housed in smaller homes and barracks. Also part of this "normal" landscape was facilities for play and entertainment. "At Buchenwald were a falconry court built as a tribute to Hermann Göring, and a riding hall for the wife of Commandant Koch. Construction of the falconry court was begun in 1938 and completed in 1940 [and included a falcon house, a hunting hall, circular garden house, and a falconer's house]. There was a game preserve and a cage for wildcats. Fallow deer, roebucks, wild boar, a mouflon, foxes, pheasants, and other animals were kept there. Outside the falconry court, in the so-called zoological garden, five monkeys and four bears were kept in cages. In the early years there was even a rhinoceros . . . one of the satanic SS pastimes under the regime of Commandant Koch was to throw prisoners in the bears' cage to be torn limb from limb. The animals enjoyed an excellent diet."[47]

The rules governing this section of the concentration camp were intended to remove one's home life from the daily activities of the camp. The residential area, with its entertainment were to present the appearance of an idyllic German community—cultured, orderly, and materially prosperous. But no matter how carefully done, erecting this surface could not erase the reality beneath it. Even the Nazi residential section was built and maintained by the slave labor of the camps. These slaves made their appearance in thinly disguised roles of servants, gardeners, and other maintenance workers. On other occasions, appearances were discarded entirely, as when the caged animals in the zoo were treated more humanely than those humans caged behind the barbed wires of the concentration camp proper. There is something brutally revealing about feeding the prisoners to the bears.

At Auschwitz, we find similar attempts at creating a pretense or a surface to disguise reality. The appearance of the Nazi residential areas was similar to those in other concentration camps. Höss, the commandant, along with his wife Hedwix and their five children, lived "in a tree-shaded stucco house known as Villa Höss. It stood just outside the northeastern corner of the camp, separated from the neighboring barracks by a concrete wall high enough so that nothing inside the camp could actually be seen by Höss's family. Near the wall, Frau Höss grew rose hedges, and begonias in blue flower boxes." Even though this landscape was clearly constructed and supported by inmates, Höss refused to admit that their relationship to them was anything but kind and generous: "My wife's garden was a paradise of flowers . . . [Höss recalls] No former prisoner [who worked as gardener or servant] can ever say that he was in any way or at any time badly treated in our house. My wife's greatest pleasure would have been to give a present to every prisoner who was in any way connected with our household. The children were perpetually begging me for cigarettes for the prisoners ... the children always kept animals in the garden, creatures the prisoners were forever bringing them. Tortoises, martens, cats, lizards. ... Their greatest joy was when Daddy [now slipping into the third person] bathed them [in the swimming pool]. He had, however, so little time for all these childish pleasures. ... ."[48] These attempts at appearances helped only the perpetrators of terror, not its victims. For them, the facade did not lesson the horrors of the camps.

Places of total terror   The camps were purpose-built places for committing evil and, as we noted in the introduction (and will expand upon in the next chapter), the geography of evil involves one or more of the following qualities: isolation and autarky; homogenization and uniformity; and constant geographic transgression and chaos. All of our discussions about Nazi Germany have been exemplifications of these principles, as they applied to both victims and victimizers. Nazi censorship and control narrowed and isolated the Germans from the rest of the world; Nazi rules made German places and projects more and more alike; and Nazi authority allowed the government to transgress and destabilize places and projects that were controlled by anyone, including German citizens. How this evil pervaded various types of German places will be discussed soon; here I want to focus on the concentration and intensity of all three qualities of evil that the camps were designed to inflict on their victims. This microgeography of condensed evil reveals both the necessity of place and yet how it ultimately interferes with evil's perfect execution.

The camps isolated and shrank the prisoners' world to nothing. Power within the camp was unimpeded and tyrannical, and no place or project could be undertaken that could not be transgressed. Yet the execution of all three qualities of evil within these small areas was often out of sync. Isolation could interfere with homogenization, which could interfere with transgression. The result was an unevenness in the landscape, allowing for tiny degrees of relief from the ultimately overpowering terror. These small degrees of nonuniformity were compounded by the fact that there was more than one form of pain to be inflicted. In the case of the Jews, the camps were to serve as the means to their death. But the concentration-death camps held many others as well, and even in the case of the Jews, the Germans wanted those not immediately gassed to die at varying rates through inhuman working and living conditions. The multiple projects of work, torture, humiliation, and death led to different geographic needs that were difficult to coordinate. And, there was yet another important complication—that of the intoxicating effect of absolute power itself.

Nazi ideology was about hate, and concentration camps allowed the Nazis to concentrate and vent it. To be able to hate and act upon it without any reservation—to have complete and unbridled control over the objects of hate—one has to have absolute power. One has to be godlike (or demonic). This experience of absolute power involves not only taking life, but also giving it back. It involves toying, like a god, with the fate of victims, sparing them at one arbitrary moment and crushing them the next; and playing god or the devil required some degree of geographic variation within the camp.

The camps were surrounded by high-voltage barbed wire and a ring of observation guard posts. These were usually arranged inside according to a rectangular grid that would provide the maximum means of surveillance. Most structures were made of wood and consisted of rectangular barracks for the inmates, with a large open area nearby that was used to assemble and often punish the inmates. This was surrounded by workshops and factory halls, gardens, mess hall, jail (which was used as a torture chamber), crematorium, occasional infirmary or hospital, and, after 1940, in the larger camps, one or more gas chambers (although these were sometimes outside the barbed wire fence and hidden from view, though themselves enclosed in barbed wire).

Terror reigned throughout the camp. These microplaces though designed for terror, made its execution less than uniform. In the filth of an uncleaned latrine, with mounds of fecal material, one is at least for the moment beyond the gaze of the officials. While capos—supervisory inmates whose lives were temporarily spared—lived in the barracks among the inmates, these were so packed with bodies that one might again temporarily escape notice. But while terror from the gaze of authorities might be stayed, a different kind of terror occurs in never being alone. Inmates were packed so tightly within the barracks, with several to a bed, that even in their emaciated state, if one needed to turn then all had to turn en masse.[49] One could not easily remove the dead, or move oneself from them. Because sanitary conditions were virtually nonexistent, latrines few and inadequate, disease was rampant, as corpses accumulated on the grounds and in the barracks.

If torture and death were the aim, then why were there infirmaries and hospitals? Death was certainly the end point, but the execution and enjoyment of absolute power meant that one could not only kill, but grant reprieves in order to toy with one's victims. And there were the all-too-infrequent and short-lived instances of kindness. Humans cannot promulgate absolute terror consistently and all of the time, for even the most evil of them cannot completely escape their humanity.

"In theory" says Friedrich, the author of the *Kingdom of Auschwitz*, "there was no reason why a death camp should have a hospital at all, and yet the one at Auschwitz grew to considerable size, with about sixty doctors and more than three hundred nurses. It had a surgical department and an operating theater, and special sections for infectious diseases, internal injuries, and dentistry. Yet the lord of this domain was Dr. Mengele, the chief physician at Birkenau, who labored long hours on testing and then killing captive twins in a futile effort to find new ways of increasing the German birth rate." A nurse, who worked with him, said after the war that he insisted on delivering a baby and took "every precaution during the accouchement . . . watching to see that all aseptic principles were rigorously observed and that the umbilical cord was cut with care. Half an hour later he sent the mother and child to the crematory oven."[50]

For anyone who had a serious disease, the favored solution was gassing, but word of this made everyone avoid the hospital and soon infectious diseases such as typhus were spreading throughout the camp and even endangered the Nazis themselves. Some of the SS simply made the rounds of the camp and would order anyone who looked like a typhus carrier to be gassed. But this did not curb the disease, and because the camp still needed to supply laborers, and also to prevent the Nazis from falling prey to these contagious diseases, the hospitals sometimes did cure the sick, but again to return the patient to the larger torture chamber—the camp. The cures here were almost entirely instrumental. It was not for the good of the person, but for the function the person fulfilled. There were however examples of acts that were more genuinely kind. Doctors could override the decisions of the guards, and so occasionally show mercy. Still, what few acts of kindness were shown, were circumscribed in their import: they did not overturn the prisoner's ultimate end.

The projects of inhuman work, torture, humiliation, and efficient and mass exterminations complicated the geography of terror. And so too did the expression to absolute power. The most vivid case where this joy of power interferes with efficient death is in the Nazis' attempts to prevent inmate suicides. Suicide should be allowed, it would seem, because it would abet the goal of destroying inmates. The reason it was prohibited was that it allowed the individual to decide how, where, and when he or she will die. That is, it demonstrates the existence of a will that is not controlled by the terrorist. Suicide becomes an act of defiance. Absolute power allows no room for the exercise of a will, not even the will to die. Death and its conditions, as well as the granting of life, must remain completely in the hands of the prison guards.[51]

Hate, and hate of Jews, was central to Nazi ideology, and so too was the worship of unbridled power, which is Fascism. Hitler's Fascism was a particularly savage one with a Social Darwinian vision of might making right. The weak and needy are thrown aside, and political institutions are not checked and balanced to prevent the abuse of power, but used to generate it. Unbridled power as an end is an intoxicant, and hate provided a direction and focus. The concentration camp was the place where hate and unbridled power merged. Still, the purposes of the camp diverged sufficiently to create a landscape that had its variations, its surfaces and depths and its places of torture and temporary reprieve. Even this intense geographic concentration of power and expression of hate did not always cover the landscape uniformly.

The Weave in Ordinary Places

We have been focusing on the sites at which Nazi ideology and state power were distilled and most concentrated—the camp. But virtually every other type of place in the Nazi landscape contributed to the development and spread of particular facets of this ideology. Each added its own particular knots to the weave of truth, justice, and the natural. To make the point I turn away from these extraordinary landscapes of the camps to places of everyday life. What then were the projects and practices in the home, the school, the church, and the workplace, as each in its own way became an instrument of Nazi anti-Semitism?

The home Home is generally among the least specialized types of places in terms of its mix of meaning, nature, and social relations. But all places in Germany were to reflect Nazi views, and thus contribute to the Nazi weave. Special emphasis was given to the natural. The form nature took for the Nazis was race, volk, and in the home there was also the biological role for women, and these "natural" elements were woven into the fabric of domestic life.

In most industrial societies home is usually a place that is juxtaposed with the site of work, and in the Nazi era, most work was particularly dirty, and so placed a special burden on home to provide the opposite, an especially pure environment. A purified and perfect and natural German world was the aim, and until the job was done, work was dirty, and the dirtier it got, the purer had to be the sanctuary of home. The most intense and geographically specific juxtaposition between the impurity of work and the attempt to purify the home is provided by the residential landscapes of the SS and guards that abutted the concentration camps. As we saw, this environment had to create not only comfort but also normalcy and peace into which the SS could escape. Höss, the Auschwitz commandant, described his home, against the wires and walls of the concentration camp, as a haven of peace and goodness. The same need was felt when the home was far away.

Associated with the image of pure home and impure work was the gendered distinction of home as the domain of women, and work as the place for men. Nazi ideology was not only racist, but also blatantly sexist. In *Mein Kampf*, Hitler said bluntly that women are inferior, and "the goal of female education must invariably be the future mother."[52] Women were to be banned from the public world; as inferior beings, they should be protected by men, and home was their place. Their duty was to make the home a nurturing environment to raise sons for the fatherland and also future mothers. Men could serve the state in many ways, but there was only one useful though essential function for women: to bear and raise children. The role of women, according to Goebbels, is to "increase the preservation of the species and the race. This alone is [their] meaning and . . . task."[53] Hitler saw the gratification that women themselves would feel for being good child-bearing and rearing Nazis as "the reward which National Socialism bestows on women in return for their labour. [They are rearing] men, real men, decent men who stand erect, who are courageous, who love honor. . . . Womanhood must say to themselves, 'what a robust and glorious generation is growing up here!'"[54] The "here" of course was a racially pure home, and soon a racially pure neighborhood, country, and world. Unlike everywhere else, which still had to be purified, the Nazi home was already cleansed of racial contaminants, and so the women who remained home were kept pure. Moreover, the home allowed mothers to instill Nazi ideals and create racially pure and socially harmonious individuals. The care and nurture provided at home were in direct contrast to the violence and destruction that were unleashed everywhere else. Home "helped Nazis to preserve their self-esteem and to continue their work under the illusion that they remained decent."[55] But there was another side. In assuming this role, women were allowed a domain that they could control. Even though their position was subservient, Nazi ideology gave them "their own realm against male interference."[56]

Women had their domain, but it was at the forbearance of the men and their conception of women. Hitler defined what women were, prescribed female roles, and provided the ground rules and authority within the home. Hitler even controlled reproduction. "Birth control was outlawed and marriage-counseling centers were closed or replaced by 'eugenic counseling' facilities."[57] Babies were a female's greatest product. Women received medals according to the number of children they bore, and when a Hitler Youth met such a woman he was to salute with a Heil Hitler. The Hitler Youth motto for girls was: "Be Faithful, Be Pure, Be German." Boys, on the other hand, should: "Live Faithfully, Fight Bravely, and Die Laughing."[58] So far did Hitler's power penetrate into the home that even a child's allegiance could be expected to reside not with the parents, but with the Führer. A child could inform on his or her parents for not being good Nazis.

The home then was the center of acculturation in which women played the pivotal role. They absorbed the anti-Semitism (and also the sexism) of the Nazi position, bringing into the home the ideas of racial purity, eugenics, and volk so that even before "the Gestapo summoned Jews to deportation centers, 'Aryan' friends and neighbors had excised Jews from society. Although the last stages of the 'Final Solution' remained entirely within Himmler's authority, women as well as men, delivered up the victims."[59] At the same time, women were to make the home into "an island of serenity where love, tenderness, and devotion reigned. A place to 'touch base' and reaffirm one's humanity in the face of brutal criminality . . ."[60] And Nazi women, though complicit in the process, did not want to be aware of what took place in the public arena. "They actively cultivated their own ignorance and cultivated [men's] escape."[61]

These female roles were traditional ones, and while they applied to all women of the Third Reich, the true home was to be, ideally, in a rural setting. An agricultural landscape would provide a counterpoint to the driftless and cosmopolitan qualities of the city and of the modern world. Rural and village life would intensify the bond between Aryan blood and German soil. Ironically, the best example of what the Nazis had in mind as a rural German paradise is found in the plans for Auschwitz—not in the camp of Auschwitz or the SS sanctuary, but in the small city and its hinterland that slave labor from the camp would help transform.[62]

The church   Nazism was hostile to any competing ideology, including Christianity, but especially early on, Hitler was careful to court the churches by drawing on a shared view of the Jews. Not surprisingly the churches for the most part harbored the same degree of anti-Semitism as the populace at large. "Never once did any German bishop, Catholic or Protestant, speak out publicly on behalf of the Jews . . ."[63] The Catholic Church did not officially protest the 1933 boycotts, the Nüremberg laws, Kristallnacht, or the Jewish deportations to the death camps. German Churches did officially protest the euthanasia killings of the mentally ill that were run by the Nazis in the late thirties and that was supposed to weed out inferior German stock, but not the gas chambers and crematoria of the concentration camps,[64] and sermons and pastoral letters from 1933 on were made in support of Nazi laws regarding the Jews. In 1936, the German episcopate announced that "race, soil, blood, and people are precious natural values which God and the Lord has created and the care of which he has entrusted to us Germans."[65] The record for the Protestant denominations is not very different. Those Jews who had converted to Christianity even before 1933 were expelled from churches.

Koonz describes the reactions of a Jewish woman who had converted to Catholicism before the Nazis came to power and who was then immediately expelled from the church: "Nothing but nothing could ever hurt as deeply as the day . . . when members of her parish had told her that 'Aryan' Catholics found it unseemly to attend Christmas services in the presence of a Jew. How they had rejoiced at her conversion so many years before, and welcomed her and her husband into their community! Then, after 1933, these same Christian neighbors no longer greeted her on the street. . . ."[66]

The tacit support of the churches is interesting in that Hitler in many public occasions denounced Christianity and especially Christian doctrines of charity, equality, and brotherhood, which he believed were fundamentally Jewish cosmopolitan ideas. The sentiment of volk and the pervasive anti-Semitism within the ranks of the churches explain their support of Hitler. But equally interesting is the attempt by some to create a Nazi church. The German Christian movement and the church they founded in that name saw Christianity and National Socialism as mutually reinforcing. This was the case because these "Christians" interpreted Christianity in an entirely Nazi light. Not surprisingly, they denounced the Old Testament. More surprisingly, they argued that Christ was not a Jew, and most incredibly they claimed that the core of the New Testament was not about salvation but about racial struggle. "The teaching of sin and grace . . . was a Jewish attitude and only inserted in the New Testament. . . . Concern with sin . . . was a Jewish element to be purged from Christianity." In their view Christ was a warrior-hero "who fought against the Jews and the Pharisees."[67] They wanted to create a church that "would provide spiritual expression to the racially pure nation" and to the dogmas of racial and Christian anti-Semitism and the glorification of the masculine."[68] This small, extreme, but influential, group made explicit what many thought were possible connections between Christian and Nazi views. Their idea of what kind of truth the Christian God revealed resonated throughout the Nazi landscape and affected what people believed about justice and the natural. (There are of course churches in the United States that espouse similar doctrines, but these are neither affiliated with major political movements or with most other religious institutions.)

The school Truth was distorted everywhere and especially in schools. Jews were purged from the teaching professions early on in the Nazi period, so that Nazi schools, like homes, could be centers of Nazi indoctrination. The Nazis determined what was to be taught at all grade levels, and also the contents of the textbooks. For Hitler, "The crown of the volkish state's entire work of education and training must be to burn the racial sense and racial feeling into the instinct and the intellect, the heart and brain of the youth entrusted to it. No boy and no girl must leave school without having been led to an ultimate realization of the necessity and essence of blood purity."[69] All knowledge had the purpose of glorifying and perpetuating the state. Hitler held that "the training of mental abilities is only secondary. And here again, first place must be taken by the development of character, especially the promotion of will power and determination, combined with the training of joy in responsibility, and only in last place comes scientific schooling."[70]; and for Hans Schemm, the chair of the National Socialist Teachers Federation, the goal of the state was "to educate men and women, who are morally, racially, ethically, and in personal character completely German."[71]

In Nazi education, the moral or ethical is entirely driven by the racial. Courses and texts in biology were to create race consciousness and develop in youth an attachment to the soil and the fatherland. The Journal of National Socialist Education in 1937 stressed the importance of using racial stereotypes in classrooms. Students must learn to see what Jews look like. "Pupils should 'experience' certain features as right, familiar and belonging to them and their people, while they were to recognize others as 'alien' and so instinctively and emotionally reject accepting them in our national community."[72]

Elementary texts taught students that "every living being belongs to a certain living space"[73] and that "people and races could only flourish in the long-term when the environment corresponded to their original living space,"[74] which of course meant that it is a law of nature for particular groups to occupy particular territories and the Germans will and ought to occupy their natural land, while the Jews, not having one, are really not like any other living creature. They also claimed that "the natural calling of woman was to be housewife and mother."[75] Jews were depicted as "the international enemy, traitors, arrogant, unsuitable for forming a state, ruthless and inhuman, devoid of creativity and true religion, the spirit of Jews as 'critical, destructive and corrosive, but never constructive,' the Jew as a plotter of murder, treachery . . . and cruelty . . . so it is no wonder that in the criminality of all ages and peoples the Jews stand at the peak."[76]

The role of biology spilled over into virtually all other subjects. The mission of geography, for instance, was to teach the idea of attachment to soil, and Lebensraum. A geography text included the studies of the home region, emphasizing the importance of love for it and for nature; love of the fatherland, national pride, self-sacrificing devotion; hostility toward large cities; and blood and soil ideology. Another included race studies, with the Nordic races as the superior race, and racial hygiene (the inferiority of mixed races), inferior races, and the Jews.[77] History was to instill pride: "We do not learn history just in order to know the past; we learn history in order to find an instructor for the future and for the continued existence of our own nationality. That is the end, and historical instruction is only the means to it."[78] Even natural sciences were to serve the ends of the state. "Science, too, must be regarded by the folkish state as an instrument for the advancement of national pride. Not only world history but all culture must be taught from this standpoint."[79]

The home, the church, the school, and every other type of place within the Nazi system still specialized in different combinations of meaning, nature, and social relations, and truth, justice, and the natural. But they were all influenced by and performed the function of contributing to the goals of the Nazi state, the perpetuation of racial purity, and the dissemination of racial hatred. By no means did all function smoothly; conflicts of means and interpretation occurred within every place, and the goals themselves were often unclear and contradictory. Even within the most singly focused place like the concentration camp, we saw that terror interfered with efficient extermination, and work as a means of torture interfered with both.

There existed much tacit if not active support by the German population for Nazi ideals. Anti-Semitism was so much a part of German culture that Hitler's virulent brand was not very much out of place; nor were the geopolitical concepts of volk. Concentration camps were to some degree public knowledge—they had to be because of their sheer number—and so too was the fact that Jews were being tortured and killed. After all, you cannot remove so many people from your midst without knowing that something was happening to them, and many Germans had relatives who were guards in the camps. However, most of the extreme horrors of ghettoization and the Final Solution were intended to be secret, for these atrocities would be revolting even for the ordinary German anti-Semites. This does not necessarily mean that Germans in general were Hitler's "willing executioners." What it does mean though is that Germans did not want to go out of their way to know, and that they knew enough to have known better. Indeed it was clear that German popular sentiment would sway Nazi officials. Germans at large were repulsed by, and through public protests helped end, the Nazi policy of euthanasia because it was directed against German citizens, albeit "defective" ones. But concern about Jews and non-Aryans was entirely another matter. Still, the worst places—the concentration-death camps—were like black holes. Something terrible was known to be taking place, but no one wanted to see through to find out, for what they might see would be so horrifying, that their own system of beliefs would crumble.

Sites of Resistance

The Nazis employed the full range of place's structure and dynamics to achieve their purposes. Those who resisted the Nazis did not control large-scale places, but in the places they did occupy, they too used the same range of dynamics for their own ends. Spatial flows could be disrupted, rules of in and out could be transgressed and changed, appearance or surface could disguise the reality of resistance, and all of these could be employed to create a different weave of elements.

In many respects, the home, the church, and even the classroom could be turned into such sites of resistance. The actions need not be heroic, but could nevertheless blunt the edges of the Nazi message. One simply need not do all that the Nazis expected. A parent may not teach her children to hate. A teacher may somehow soften the message of an anti-Semitic text, and the clergy could emphasize the Christian virtues of charity and redemption. Even such small gestures would make such places more open and allow greater variety and complexity. And then there were the more explicit forms of resistance, many of which produced virtually unparalleled acts of heroism.

The German underground, though weak, undertook some forms of sabotage and guerilla warfare, and there was active resistance among the victims themselves, as in the Warsaw Ghetto uprising. These important forms of struggle were ways of transgressing the geographic power of the Nazi state by violating the rules, by moving things where they did not belong, and challenging the meaning of what was taking place. Although the defeat of Nazism required the intervention of the Allied powers, these internal forms of resistance were extremely important.

Another form of internal resistance was nonviolent, but active and extremely heroic. There are two strands to it. On the one hand, Jews in the ghettos tried to carve out even a tiny niche that could provide them with some small respite from terror and that could even be used to create something culturally positive: a place that would not only help sustain the body, but allow some form of social relations and meaning to continue. This could hardly be done in the concentration camps and was barely possible in the ghettos. Still, they tried. In the ghettos, meaning was sustained to a degree through clandestine prayer meetings, secret houses of study or yeshivot,[80] underground libraries, and the development of secret centers of learning "in back rooms, on long benches near a table."[81] Even the act of keeping a diary was a form of resistance. The uniformity and intensity of terror in concentration camps was so strong that even these small "para-places" were not available. One could retreat only into one's body. Yet, even here, there were possible connections to a world outside. Some were occasionally able to read (parts of) books because a kinder or thoughtless commandant might issue to the camp the pages of discarded books as a substitute for toilet paper.

The second strand involved the acts of non-Jews who harbored Jews at the risk of their own lives and the lives of their families. This meant that a room, a closet, a root cellar, or an attic was put aside as a place of refuge, for weeks, months, and even years, and in some rare cases entire villages harbored holocaust victims. How did these places work? Clearly, they were in violation of the rules that pertained to the society as a whole. Those who provided them were criminals and would be executed.

Still, there were those who did. Heroic Germans include those who sheltered and protected the approximately 2,000 Jews who lived out the war in Berlin alone (compared with the 170,000 pre-war Jewish population). Some have estimated that these 2,000 needed the help of between 20,000 and 30,000 Berlin citizens.[82] Those who helped not only faced death, but in their daily lives had to steal and lie in order to keep the Jews safe. In other words, in addition to defying Nazi edicts, they also violated moral codes and covenants about lying and stealing. A particularly poignant example of rescuers was a Dutch couple. They were of modest means and lived adjacent to their pharmacy in a small community. They built a secret room in their pharmacy, big enough for ten people to stand. All told they hid 37 Jews, while raising five of their own children and one of a neighbor's. They were soon betrayed. On a tip from a neighbor, the Nazis searched the store and house but were never able to find the secret room, and so the rescue continued. During the war the wife contracted tuberculosis and was dying. The people they were harboring offered to leave because they believed she contracted the disease from them, and that they were causing too great a burden and were posing too much of a threat. The husband thought that a solution might be to have his wife live with her relatives in another town. She refused to leave and also refused to have the Jews leave. In the end they were saved, but the wife died from her ailment.

When asked why such heroes (whom the Jews call the Righteous Gentiles) did what they did—defying the mores and ethics of their own communities, their neighbors and friends, and risking not only their own lives but the lives of their loved ones—the almost universal reply was that they did not see what they did as in any way out of the ordinary. They did what any human being had to do. It was simply the right thing; they could not have done otherwise.[83]

Another case, involving virtually an entire village, is that of Le Chambon, a place of approximately 3,000 poor people in the Department of Haute-Loire in south central France who, at great risk to themselves, provided sanctuary to about 5,000 refugees, most of them children.[84] This was a close-knit, predominantly Huguenot community with a long-held commitment to pacifism and with an inspired and dedicated minister who throughout the war continued to urge the community never to turn anyone away. Differences among Jews, Catholics, Protestants, or any other divisions were entirely superficial. Every human being deserved equal care and respect. This view was held by almost every single individual of the village.

Le Chambon was located in Vichy, France, an area that was ruled by the Nazis through their puppet Vichy government until 1943, when it was ruled directly by Germany. Any and every place within the community was under Nazi control, and so, as in the other cases, offering asylum meant doing so mostly in secret. The community was divided into cells so that few knew the entire network, discussions and coordination occurred in the relative security of kitchens, refugees were kept in homes, and when the village was raided by the Nazis and their representatives, refugees were either secreted in special places or evacuated to outlying areas.

Rooms, attics, closets, and root cellars were the geographical scales of most forms of Nazi resistance, and while an entire monastery, or even a village like Le Chambon,[84] could be a haven, it too contracted into these tiny geographical scales when the Nazis entered the community. These places of resistance had to violate or transgress the normal Nazi view of truth, justice, and the natural to do what was good. They had to be secretive. At a superficial level, the structure and dynamics of sites of resistance were similar to the sites of terror. The boundaries and the appearance of the places had to be opaque enough to not allow the enemy to see through, and rules and flows had to be rigidly enforced. Indeed, offering someone refuge, though an heroic act, requires that the guest be subject to the strict rules and regulations of the host.[86] It is, in a sense, an altruistic gift given with enormous strings attached. It works only if the refugees follow instructions.

This use of place again may appear to be, like evil itself, a contraction of the world. But it is not. Those engaged did so willingly, so the place was not one of oppression, but of freedom. Those who participated were motivated by an altruistic, open, and generous impulse toward all human beings. The activities of these rescuers fly in the face of self-interest. As Philip Hallie said of the villagers in Le Chambon: "by resisting a power far greater than their own they put their village in grave danger of massacre." They did so because they followed their conscience which meant "refusing to hate or kill any human being" and never turning away someone who needs help.[87] Even the need for secrecy was tempered as much as possible. In the case of Le Chambon, the villagers were as candid with the authorities as they could be "without betraying the refugees." When asked, they admitted there were refugees there, but they were not willing to disclose where in the village, nor were they willing to disclose names. They combined "candor and concealment . . . , a yearning for truth and . . . a commitment to secrecy."[88] They even felt guilty about having their children practice deception, and worried about how they would "unlearn lying after the war . . . or [if they would ever] again be able to understand the importance of simply telling the truth."[89] So even though their places were deceptive and secret, they were also places that enabled them to resist the homogenizing and narrowing forces of evil. By resisting, they created a landscape that was more varied and complex, that enabled people with different practices and views to coexist, and that encouraged an expanded and clearer view of reality. They could of course have done so with far greater openness and success if they were not surrounded by the evils of absolutism. But even the good must use some of the strategies of evil to resist.

This of course is not the way instrumental judgments would view it. The relativist side would see these sites of resistance as simply a different project, but not a better one—that what the virtues meant in the attic, closet, or root cellar, or even in the entire town of Le Chambon, were not morally superior to those of the concentration camp; they were simply different. And the absolutistic quality of the instrumental judgments of the Nazis would have them condemn these additions to the landscape because they increase awareness and add variety and complexity.

How was it possible to provide and maintain an expanded view, and yet still keep what took place secret? How did the different sites of resistance combine the virtues differently? These and a host of other questions about the geography of evil and sites of resistance need to be pursued to do justice to the problem, but to do this here would have us lose sight of the intention of this case study: to provide enough of an example of these dynamics that would allow us to imagine how this kind of analysis can be done for other regimes. With this understanding, I turn now briefly to sketch geographical evils presented by the Soviet Union and the system of slavery in the antebellum American South.

THE SOVIET UNION AND SLAVERY IN THE ANTEBELLUM SOUTH

The Soviet Union was formed before the rise of Nazi Germany. It too was a totalitarian regime whose geography of terror was similar to that of the Nazis, and several of its practices served as models for the Nazis. Even though we find a similar use of place in the context of control, the Soviet Union and Nazi Germany draw on different traditions. Few moral theorists would see any of the Nazi ideals as worthy of perfecting. Yet many would be sympathetic to the core values of Marxism. At the most elementary level, Marxists uphold a basic and admirable tenet: The meek, the poor, the unfortunate are all worthy, and indeed it is they who shall inherit the earth. Moreover, Marxist utopian thought envisions a world of altruism and gift giving where "from each according to his ability, to each according to his need" becomes the governing principle. These are impulses that many would find worth achieving. The problem in the case of the Soviet Union was largely about the means used for these ends.

The difficulty lies, as we have seen in the previous chapter, in the fact that Marx does not provide a moral theory to promote these ideas. He does not rely enough on the capacity of individuals to use their reason and will to see the value of these goals and achieve them. Rather he stresses the situated quality of moral positions and how changes in social relations will produce moral beings. The good can be engineered. And there should be those allowed to be the engineers— a dictatorship of the proletariat. Whenever a theory diminishes the capacity of reason and will, and stresses instead the role of causal relations, the door is open to absolutism.

The Soviet Union was the first totalitarian state, having a single party (which was not intent on increasing its numbers but rather on recruiting only the most devoted followers), a monopoly on force through control over the military and the police, a program of national censorship, a reign of terror, and the establishment of camps (the gulags in fact began in the 1920s and preceded Germany's experiment by almost a decade) and an eliminationist policy toward the enemies of the party or the state. Unlike Nazi Germany, the enemies of the Soviet Union were not ethnic or religious groups (or races); rather the enemies were particular classes and individuals who held (or were said to hold) opinions that differed from those of the party. As a Marxist-inspired system, the Soviet Union would naturally focus on economic class structure as the major problem confronting the attainment of a socially just society and would draw attention especially to the interests of the propertied and wealthy. The Soviet state also found it useful to single out class as the enemy because it provided an easy target or scapegoat for a society that had a very simple class structure—a very few who had wealth, a small middle class, and a huge mass of poor peasants—but that had so many ethnic and cultural groups that it would be difficult to choose a villain from among them.

Terror was an essential means of control in both the Soviet Union and in Nazi Germany. But it may not be an exaggeration to say that, even though the ideals of Marxism had infinitely more moral worth than the so-called ideals of Nazism, terror was even more pervasive in the former. This is because Germans accepted Hitler and his ideals far more readily than the people of the Soviet Union accepted those of their leaders.

The Soviet strategy for removing these enemies and for controlling the rest of the population inspired the Nazis when they came to power. In the Soviet Union, the entire country was to be under a single and uniform source of state power, which took the form of terror when directed toward its enemies. Lenin and Trotsky "expressed contempt for socialists who could not understand the necessity of terror. 'If we are not ready to shoot a saboteur and white guardist, what sort of revolution is that?' admonished Lenin."[90] The formation of a secret police began immediately after the revolution, and by 1927 its powers were sweeping. In a decree Stalin demanded that "persons propagating opposition views be regarded as dangerous accomplices of the external and internal enemies of the Soviet Union and that such persons be sentenced as 'spies' by administrative decree . . . that a widely ramified network of agents be organized by the [security police] . . . with the task of seeking out hostile elements within the government, all the way to its top, and within the party, including the leading bodies of the party. . . . Everyone who arouses the slightest suspicion should be removed."[91]

Exercising this intensity of domestic power requires as impermeable a territorial boundary as possible. As early as 1917, Soviet isolation and censorship began. By 1922 with the establishment of the

Censorship Board, the state had complete control of news and the printed word, and similar control was established over all other forms of expression. Truth was in the hands of the state; the arts, sciences, and the educational system were to serve government ends, and religion was suppressed. Nature and the natural became little more than a component of the forces and relations of production. Justice too was an instrument of the state, and could not be used to check its powers. Except for processes of adjudicating minor day-to-day misdemeanors, the law, operating without a presumption of innocence and without clear standards of evidence, was a political tool and the courts a virtual tribunal meting out state power. Terror and law were interthreaded. As Lenin put it: "the court must not eliminate terror, to promise that would be self-deception or deception."[92]

Terror could appear anywhere, even in the home, where children were encouraged to report anticommunist tendencies of their parents and friends.[93] When an enemy of the state was identified, he or she quickly disappeared. This was the case even within the Communist party: "In the midst of a meeting, a man goes to the bathroom and doesn't come back. That way it is less conspicuous. Where he ended up, nobody of course asks."[94] But they did know. The Soviet camps or gulags were not well-kept secrets. Unlike the Nazi system, the gulags were not intended as killing camps and did not contain the equivalent of gas chambers. But they did create inhuman conditions. Prisoners were overworked and starved, and millions died.

A deeper understanding of these instrumental practices in the Soviet case would require that we examine, as we did in the Nazi example, how multiple types and scales of places were employed by the state to weave particular and complementary patterns of truth, justice, and the natural. This would involve an analysis of how these were fashioned in homes, schools, churches, theaters, factories, communes and cooperative farms, municipalities, and even for the Soviet territory as a whole; how the authoritarian component required not only a common weave but also one that obscured the structure and dynamics of these places; and yet how the vastness and complexity of the Soviet Union and its competing interests created dissonances and resistance to these efforts. Such an analysis would disclose that at all scales there was a tension between using place to create a communist landscape, and using it to further the interests of ethnic groups and nationalist Russians, and it would reveal the tensions between the goals of communism and those of authoritarianism. Although there is not enough time here to pursue these in any depth, and perhaps no need given the overall example of the Nazi case serving as a template for the role of place in the execution of evil, still, one small-scale example can serve to suggest how place in this totalitarian system raised different paradoxes than in the Nazi case. The example concerns the creation of public place—a most important concept in a communist society that seeks to promote public ownership and control, and yet an impossibility in a totalitarian regime such as the Soviet Union.

As Argenbright has shown through an application of my model of place, the public areas of Moscow, its streets and squares, restaurants and offices were creations of, and controlled by, the state. The state dominated the rules of in and out. "The state determined the location of factories, grocery stores, research institutes, and everything else. Housing was assigned, as was the right to reside in Moscow. . . . Spatial interaction was likewise under tight control. The movement of goods was determined by the state's economic plan. Efficient mass transit moved most Muscovites from place to place . . . and everywhere were symbols of state power. . . . virtually everyplace had a narrowly defined significance that was supposed to replace . . . [all other] interpretations."[95] The meaning of the landscape, its surface and appearance, was then also controlled by the state.

This means that while Muscovites "were in public in the sense that they were surrounded by strangers and were visible to all," it was not a public place that was theirs in any sense.[96] They had no control over its rules, flows, or meanings. In a society that was designed to eliminate alienation through the implementation of communism, these very places were sites of alienation. They did not "belong" to the people, but rather to "them"—the representatives of state power, or to the state itself. The primary place where individuals were able to regain some measure of control, where they could escape the geography of alienation, was the home and especially the kitchen. "Only there did most people feel comfortable."[97] If the state were to control even this, then there would be no capacity to form a self. But as in the example of the concentration camp, terror is never geographically uniform and complete. There are always cracks and crevices and the possibility for sites of resistance.

The Soviet Union and Nazi Germany were both totalitarian regimes. The Soviet Union singled out the wealthy classes and political enemies as their primary targets. In the Nazi case, race was the issue, although race really meant nationality and religion. In both cases, those victims who were not murdered immediately were enslaved within the camp systems. Slave labor was economically useful, but in both countries its economic value always was secondary to its role as a form of punishment, torture, and in the Nazi case, death. This was not the case in the antebellum period in the United States. Slavery there was both racist and economically significant; and while the regime was unjust and cruel, it was not nearly as centralized as were conditions in Russia and Germany. What then are some of the distinctive qualities of the role of place in a slave system in general, and in one that was not very centralized?

Slavery does not need to be based on racism; indeed it has not been for most of its institutional life which is far longer one than the relatively new phenomenon of racism. Slavery served as the backbone of economies in many places. But a slave economy is a difficult one to sustain. Even though slaves are a form of capital, keeping people as slaves and having slaves reproduce themselves require not only an expenditure in surveillance and control, but more importantly, they require that at any moment a significant part of the slave population be unproductive in that women who are pregnant or rearing young children would have to leave the workforce while they and their charges consume food and other resources. A more efficient slave system would rely on capturing and enslaving people who have been born and raised in their own local communities. Then the expense of bearing and raising them has already been borne by that community, and the slave owners will then have a ready-made productive slave. True, slave owners will now have to bear the cost of capturing and transporting the slave, but with sufficient volume, this is less expensive than raising one's own. Conquest then was always an important part of the slave economy,[98] and slavery often produced a far-flung geographic web.

Finding a consistent moral theory that would condone slavery is also difficult. The most common argument is to equate a slave with an animal, but no matter how detailed the argument becomes, it eventually runs into the insurmountable and undeniable fact that animals serving us as beasts of burden are not slaves, and that slaves are not animals. Slaveholders find slaves useful (and slavery difficult) precisely because it is humans, not animals, that are being subjugated. This fact pervades all slave cultures, for the rules and expectations concerning slaves are different from those that apply to that society's animals.

If thinking of slaves as animals does not work, then a subtler justification of slavery turns on the argument that slaves are inferior people, born to be slaves; and a corollary to this, drawing on a variant of paternalism, is that enslavement can even improve these people by introducing them to a higher culture. This idea of paternalism is in fact the closest apologists for slavery can come to a moral claim. It is found in ancient Greek society, which depended on slavery and which at least to its credit believed slavery needed justification. Those who used this idea to support slavery could then think of the institution as a beneficent one, helping both master and slave, not unlike a father's relationship to his children.

Justifications of slavery were formulated well before the development of racism, and argued for the enslavement of individuals because of their cultural or social level in a hierarchy rather than because of skin pigmentation or other physical features. But by the 18th century, racism became a strong movement, and its affinity to slavery is obvious: racism singles out one type of people as necessarily inferior and so it is only natural that they be used as slaves. But in the 18th century it was still unclear what caused this alleged racial inferiority.

Most of the eminent 18th century American and European thinkers who were racist were so with a lower rather than uppercase "r" because they believed or at least held out the possibility that racial differences were not necessarily inherent, but due to the place and the society from which individuals came. The "backwardness" of a culture, or a particularly oppressive natural environment, may make its people slave-like; if they were removed from this environment, there was the possibility they could develop. So enslaving them was justified because of the way things happened to be at the time, but that does not mean that they were to be permanently enslaved by their "race," or that their race should be permanently enslaved. Slaves and their children might well improve once they were removed from their own debilitating cultural or physical environment. Even the overt racial differences such as skin pigmentation might disappear with a change in environment, for the color of the skin and other physical features could be a reaction to a particular environmental state of affairs. But by the 19th century there was a move to racism with an uppercase "R," as most North American racists came to hold the view that these qualities of inferiority stemmed from physical features that could not be shed by simply moving individuals to new places or could not be overridden by having them exposed to new social contexts. Race then was an immutable and inheritable characteristic. This of course was a useful idea for slaveholders, for it meant that they could retain their slaves and their slaves' descendants. It also led to arguments for racial purity. Races should be apart and distinct. It was also useful to the large numbers of poor and ignorant whites who no longer would be lumped together with blacks.

Both lower- and uppercase racism were combined in the antebellum South with the social model of paternalistic aristocratic society whose privileged members were to care for those inferior members in their service. To many large plantation owners, the plantation was seen as the equivalent of the medieval manor or estate, in the sense that it was supposed to be a virtually self-sufficient political unit with the plantation owner or the "lord" of the estate at its head, and the slaves or peasants as its subjects. The geography of the plantation was the center of slave society, a symbol of what the South meant. Even though most white Southerners owned few slaves (and their own homes were not much grander than those of the slaves), many slaves were in urban centers, and slavery took place also in the North until the mid-19th century. The plantation was only a part, though the biggest and most symbolically potent one, of a complex web of places—from Africa to the Americas—creating and sustaining the Southern slave society.

No matter how much one attempts to call slavery a form of paternalism, it is at bottom an inescapably brutal socioeconomic system, and the slave is a valuable but willful and potentially disruptive piece of capital/labor. Because it is involuntary servitude, slavery, of necessity, requires that slaves be under constant supervision; that their movements be highly regulated; that their social relations be controlled; that their intellectual life be stunted (for reasons we will discuss shortly); and that their role as an economic commodity be constantly borne in mind. All of these requirements must be supported by the legal system of the slaveholding society. Slave laws throughout the antebellum South were very much alike on these matters because each state copied the codes of another and because the institution of slavery had these similar requirements everywhere.[99]

The slave's geographical place or location, like that of any other property, was entirely under the discretion of the slave owner. Within the plantation the master assigned the slave a place to live and work. In the larger plantations some lived in the great house as domestic slaves; most though were field slaves who lived in the slave quarters elsewhere on the plantation. The slaves were owned as individual units or persons and so a family was often broken apart by the sale of a mother or a father to another plantation. The slave was anchored to the site and not allowed to leave without a pass; escape was a serious offense (though one of the strongest motivations for running away was to see one's family). Slaves on farms and plantations were to be under the supervision of resident white men and not slave foremen.[100] Supplemental rules applied to slaves in cities, prohibiting them from being in streets after curfew or living apart from their masters.[101]

The slaves were not only to be restricted in terms of their movements through space, but also in their contacts with the outside world. The most important means of narrowing their world was to keep them illiterate. Books and newspapers would contain opinions that could subvert slavery, and keeping slaves docile was an important goal of discipline. In most states, no person, not even the slave's master, was to teach a slave to read or write.

Censorship limited the world of blacks in terms of what they could hear and read, and of course in what they could say. But whites too were censored. In 1836, Virginia passed a law punishing anyone who introduced abolitionist literature into the state or who spoke about abolitionist ideas. "In 1837 the Missouri legislature passed an act 'to prohibit the publication, circulation and promulgation of abolitionist doctrine.'" In Virginia in 1849, a fine and imprisonment were imposed on any person who espoused the abolitionist view "that owners have no right of property in their slaves."[102] Louisiana made it a capital offense to use "language in any public discourse, from the bar, the bench, the stage, the pulpit, or in any place whatsoever, that might produce insubordination among the slaves." Part of this form of censorship was to prevent these ideas from undermining the conviction about slavery among the whites, but part was also due to the fact that many blacks had learned to read in spite of laws against black literacy. In a petition to the South Carolina legislature, it was stated that "the ability to read exists on probably every plantation in the state, and it is utterly impossible for even the masters to prevent this . . ."[103] Maryland's censorship law may have been the most sweeping. It considered it a "high offense [punishable by a prison term of between 10 and 20 years] against the supremacy of this state for a person to write or circulate any publication, having a tendency to create discontent among . . . the people of color of the state."[104]

College books and curricula were subject to censorship. By concentrating on the classics, Southern colleges could find arguments among the Ancients in support of slavery, and avoid contemporary literature and debates against it. To assure that professors would not introduce subversive material, colleges prevented the faculty from selecting their text books, forbade them to speak their minds about slavery or politics, or even to hold unorthodox religious views.[105] Attempts were made to expurgate such dangerous ideas from the books used in colleges and schools, and landowners were dissuaded from sending their sons to colleges in the North, for fear these young men would become infected with radical ideas.[106] In 1856, Bishop Polk of the Episcopal Church founded a college in Sewanee, Tennessee that was dedicated to teaching the sound doctrine of slavery: "Here at Sewanee the sons of Southern planters could drink pure and invigorating draughts from unpolluted fountains."[107]

Since the primary objective of slavery was not extermination but extraction of labor (and also the maintenance of the social system that depended on it), the slave was to be fed and housed in a way that would not overly diminish his or her productivity and, as the slave trade began to recede, the need to have slaves reproduce allowed them a bit more latitude and independence. Field hands were watched closely during the day, but at night, they retired to their cabins where they were not closely supervised. Those that had families then could create a small and precarious place, but a place nonetheless, for domestic life. Even the slaves in the plantation house were given quarters in which they were not overly supervised, and were allowed to work in rooms unattended.

But slaves did more than use the places allotted in a way that would conform to the rules of the plantation. They turned such places into sites of resistance and into hearths of slave culture. There slaves revolted, slaves escaped, slaves did not cooperate, and slaves resisted also in a way that unfortunately is underemphasized in many history books—they taught themselves to read and write in defiance of white authorities who would severely punish such acts. Reading and writing of course opened up a wider world of meaning: it allowed the slave to vault over the restrictively bounded confines of the plantation, to see that there were other worlds and other forms of right and wrong.[108]

The result of these acts of resistance was that the very places that enforced slavery also became sites that allowed the production of slave culture. This could occur because the slaves resisted and took some control of their lives, and because the power relations of slavery were such that the spread of force and oppression was not uniform over space. There were small places, nooks, and crevices in this surface of oppression that allowed positive qualities and contributions to grow. The fact that slaves were able to produce culture raises a difficult and embarrassing problem, which our other examples raise too. It is the age-old one of good arising from evil. Heroic acts are required to counter evil, and paradoxically a good society is one which makes such extreme examples of goodness unnecessary.

Nazi Germany, the Soviet Union, and the Antebellum South used familiar places and created new ones to undertake their respective projects. From an instrumental viewpoint they met with varying degrees of success. It is difficult to imagine how slavery in a largely agrarian economy could operate without something like the plantation system and its attendant places. Certainly, they could have been made more efficient, and the legal and cultural system more effective, but overall they were effective enough. It is equally difficult to imagine how the Nazis could have accomplished their goals without a totalitarian state and something like the camp system. It is more difficult to say that the goals underlying communism had to be met the way they were in the Soviet Union. Perhaps this had much more to do with the goals of the leadership than with the core of the ideology. In any case, each of these spatial systems reworked the meanings and mixes of truth, justice, and the natural. From the instrumental point of view of the Nazis, of the many of the members of the Communist party, and of the slaveholders, their places were not only effective but good in that they helped move the society toward desired and valued goals. The instrumental position could take issue with the details of geography—for example, greater efficiency in ridding the world of non-Aryans could have been realized by turning more of the concentration camps into death camps; the gulags could have been organized to more effectively use the labor of its inmates; and slave plantations could have been more effective economic systems if plantation owners were not so preoccupied with conspicuous consumption. But this does not change the overall point that instrumentalism could be used to justify racism, fascism, and totalitarianism, and that instrumentalism alone cannot launch an attack against them. For that we must turn to intrinsic judgments, which is the subject of the next section.

II

Intrinsic Judgments

5

THE THEORY

INSTRUMENTAL JUDGMENTS CLAIM THAT MORALITY, the "ought," is a product of empirical conditions and self-interest. We have seen how this leads to either no sense of the good—a relativism—or to a misconstrued one that becomes concrete, literal, and absolute.

Intrinsic judgments make virtually the opposite claim—the empirical can be animated by the "ought." This reversal, along with what it entails about the real and the good, help us avoid the dangers of relativism and absolutism.

The general idea of the theory has already been sketched, especially in Chapter 1, and so the job of this chapter (and the next two) is to expand upon key elements of the theory. This chapter does so in four parts. The first is devoted to a more detailed analysis of the two criteria of intrinsic judgments—"seeing through to the real" and "variety and complexity"—and offers two related justifications for having them be the core of the moral theory. The first is that they make qualities of the good more accessible, and provide checks and balances that help us resist morally mistaken actions. (I also point out how their lack creates places of evil.) The second is that their focus on the role of awareness and free will captures essential qualities of being morally responsible and human.

But if free will and awareness are so important, why are they needed if the good is so compelling? Wouldn't we be "forced" by the good to be good? And if the good is compelling, then how is it we so often make the wrong choice? These problems are the subject of the second part. It restates the problem in terms of two famous and powerful images: the Garden of Eden and Plato's allegory of the cave. The first focuses on the need to choose, and the second, on the compelling quality of the good. How then are these to be reconciled?

The third part offers an outline of a solution. The key is in the process of self-deception. I propose that self-deception helps explain how doing the wrong thing lies not in willfully rejecting the good, but rather in willfully not making sufficient effort to be aware. This issue of self-deception will be explored in greater detail in the next chapter, where I use it to reexamine the examples of evil that were encountered in our previous discussions.

Not only do intrinsic judgments bring these relations together and reduce the likelihood of self-deception, but they also, as will be discussed in the fourth section, lead to an emphasis on altruism, which rescues the moral components of place from relativism.

These are the parts of the theory's relationship that must be elaborated. They, and especially the last two, will prepare us to discuss the implication of the theory for a geopsychology and a geosociology.

INTRINSIC JUDGMENTS

Intrinsic judgments make qualities of the good more accessible. This is their most direct justification. And, by doing so, intrinsic judgments serve as both a means of assessing what we do and as a guide to what we should do. Seeing through to the real, and variety and complexity are its two criteria, and these must be applied jointly. Both are equally important and each balances the other, but I have to explain them one at a time, and I will begin with seeing through to the real.

Seeing Through to the Real

To be or do good it is vital that we have a heightened and expanded awareness of reality (of which the good is a part). In geographic terms, this means it is good to create places that helps us see the world as clearly and deeply as possible and understand how the places we are constructing make up the world. As a shorthand, I have called the geographic form of this the quality of *seeing through to the real*, and it is this quality that becomes one of the bases of intrinsic geographic judgment.

Clearly this value is linked directly to the assumption that good is real and that evil is a result of a lack of awareness. The obligation to have a heightened awareness applies to us as individuals and to us as part of social and cultural structures and institutions. Being involved in large-scale projects over which we have little control may make it more difficult for us to understand the consequences of our actions, but it does not absolve us of the responsibility of being aware. Seeing through to the real is an obligation that not only applies to us personally in that each of us should see as clearly as possible, but it is also a collective responsibility. We should help others have the opportunity to develop their views and allows all to share. Awareness is stifled and endangered if it is kept private and secret. It must be free and open to the public, and given as a gift. Making awareness public allows our views to be tested against others and clarifies the picture for all.

It is not surprising then that valuing an awareness of reality is a deeply held part of our human nature. It draws attention to our intellectual capacities to reason and to pursue truth. In geography it is expressed as a curiosity about the world and wanting to know what lies beyond the horizon. This curiosity is manifested in the cartography of every culture in every period, from the ancient and preliterate to the modern, and how each culture's maps provide a picture of the whole, even when it is a mythologized picture of the world and the cosmos. Seeing how the world and its multiple parts interact—seeing through to the real—leads to good effects. A heightened awareness of the real increases our understanding of the consequences of our actions, which we must have in mind if we are to be morally responsible agents.

But, one might say, even if this is an important value in geography, is it not again an instrumental one, necessary for geography and for any project? It is true that any project must enable us to see how things are connected; therefore, a greater awareness of the real increases the effectiveness of our own projects and seeing through to the real then could be part of any instrumental set of values. But most often the real that we are interested in when engaged in an instrumental project is defined and constrained by the scope and purpose of that project. Seeing the real from an instrumental point of view also makes it likely that we will keep our knowledge to ourselves. Revealing our ideas about the real to the public may in fact run counter to the project's goal which may thrive on detailed and confidential knowledge. In contrast, the value of seeing through to the real that is encapsulated in intrinsic judgments is a heightening of public awareness of the real— an expansive awareness that examines publicly the whole world and its parts.

Again, one might object that seeing through to the real is still part of a project, in this case, the project of geography, and thus remains instrumental. This objection is countered when we remind ourselves that this value is a deeply human one, and can be part of any act. We are intellectually curious beings, insatiably so. We not only want to see reality, we probe it, using our imaginations to understand what lies behind or beneath what we see to create visions of new realities which become realized as part of our future projects. We do so not only for practical reasons but also simply because of the magnetic pull of the real and the good.

Taking seriously this connection of seeing through and the magnetic quality of the real and the good, sheds light on the geographic problematic. We can interpret it as saying first that we are incapable of accepting reality, not as it is, but *as it appears*; we wish instead to know it *as it really is*. Second, since we are not only observing reality, but also changing it through our place-making, we can add that these changes are made in order to improve upon reality—*to make it better*. If we have a true—that is a morally correct—understanding of what it ought to be, and act upon it, then these two interpretations coincide. But all too often these images of an improved reality are pure fantasy; so a renewed pursuit of the real is a necessary corrective to make sure that we do not make disastrous mistakes. Of course, we can never know reality *truly as it is or in its entirety*, and of course we are changing reality by creating place with at best only an intimation of the good; yet the success of these changes depends on how well they conform to a deeper understanding of what is real—deeper in that our changes are realistic in that they conform to principles of the natural and human world and do not diminish this world to the point where it limits our life chances and opportunities, and deeper in its understanding of the good.

An expanded awareness that comes from seeing through to the real is then something that is good and that also has good effects. If we are interested in doing the right thing, we would want to know more about the real, for this will allow us to understand the consequences of our actions. Awareness of our effects are even more important now that we are geographical leviathans radically transforming the earth to the point that our actions affect even the possibility of our own collective capacity to survive, and hence awareness itself. At this point, awareness of the consequences of our actions becomes its own precondition.

We can say at this point that a good place then is one that enhances this capacity of seeing through, not simply for those within, but for everyone. It means that its boundaries must be permeable or transparent enough to allow the outside world in, to allow those within to see beyond, and to do so with as little distortion as possible. But how can a place, that, by definition, is bounded by rules and that contains only some things and excludes virtually all else, enhance our vision, and encourage us to have an expansive view? Here I will tackle part of this difficult problem and leave the rest for Chapter 6 where it is taken up as an issue of compartmentalization. It might help if we remind ourselves that we are assuming such places can exist when we talk about cosmopolitan, worldly, and democratic places, as opposed to provincial, stifling, and dictatorial ones. These contrasts suggest what is meant. But perhaps the university, as a place dedicated to pursuing truth and understanding reality, is really the archetypical case that, when functioning as it should, expands our awareness—not narrows it. The university makes its knowledge public. Lecture notes are allowed to be circulated, and professors often publish. Both teaching and publishing are described as contributions to their fields, and this contribution is also part of the public domain. It is the place that should, in principle, be open to public debate about what is true and real.

Holding then the university in mind as an ideal of a place striving to help us "see through," let us consider a more complex and larger-scale place that can be said in many respects to illustrate the opposite—a place that prevented people from being aware. Of the many examples raised in Chapter 4, perhaps the Iron Curtain of the Soviet Union stands out as a particularly vivid one. Socialism or communism have admirable moral foundations, and it is important to have people who are committed to them be provided a place to live according to these principles (and thereby increase the world's variety and complexity and thus conform to the second criterion of intrinsic judgment). But the role of the Iron Curtain in not allowing individuals to "see through" so corroded the values of this way of life that what variety and complexity resulted from the Soviet case does not compensate for or override the absence of seeing through. In this way, the Iron Curtain, although part of reality, and a part that expanded the real, nonetheless was anathema to the good of understanding the earth and its parts, and the consequences of our actions. A portion of geographical reality was then interfering with our ability to understand it and the rest of reality, and so, for this case, the place fails so miserably in meeting the intrinsic judgment criterion of seeing through to the real that this alone is likely to be sufficient to condemn it.

The Iron Curtain is an historically important and extreme example of removing a large section of reality, but to return to our general question: Does our criterion for judging condemn any place that to some degree removes what takes place from view? Part of the difficulty in addressing it is that the capacity of place to obscure involves more than the impermeability of its boundaries. Obscuring and narrowing of vision can also come from a place whose landscape—or surface or appearance—hides the reality beneath. That is, the very fact that places create a weave can obscure what is really happening as much or even more than can its boundaries. This claim can be made against any place (as we noted in Chapter 2), and so too can it be made against the Soviet Union as it often appeared to be what it was not. Surface/depth issues then are as general and important as are boundary issues, and must be borne in mind when evaluating place. But to simplify matters I will continue to concentrate on the role of the boundary.

To return to the question: Should all tendencies of place (whether through boundaries or issues of surface/depth) to narrow or obscure our vision be condemned? If so, then we would be condemning all places, for place must obscure to some degree. It bounds and compartmentalizes things, and affects what takes place, and in so doing, must narrow what can be seen. This is in fact how it helps us undertake projects. At its best, it allows us to focus on something that is most often within the place—a project, a task, an undertaking—which one hopes can then provide insights into part of reality. But only a part. And to have undertaken the project means we have blocked out the rest of reality. This dilemma can be put most directly in experiential terms. In order to contribute and provide awareness to others, we need to withdraw into the confines of a place in order to be aware, to focus, and create. So even at its best, place simultaneously contributes to a heightened awareness of some parts, while ignoring others (which, as we shall see, is one reason we need the second criterion that is met by a variety of places). Clearly, a balance must be found between duration and intensity of the withdrawal and isolation, versus the openness and engagements.

Another problem arises when we consider balancing the views of those within with those outside. Those within who are engaged with the project know what is taking place, while those on the outside must simply accept that what is happening will ultimately be justifiable. Even so, it can create suspicion for those outside; and for those inside, being to some degree immune from public gaze, it can lead to abuses. The difficulties in establishing a balance are highlighted by the fact that the first criterion of intrinsic judgment urges us to create place that provide the opportunities to think about and understand the world, and these same places become a significant part of the world we wish to explore.

To help establish the balance between transparency and opacity of place (expressed either as an issue of boundaries or as an issue of surface/depth), we must appeal to the second criterion of intrinsic geo-graphic judgment. But, before I do, even the "seeing through" principle and its requirement that knowledge be shared, along with the following caveat, will go some way toward establishing a balance. The caveat is that the opacity necessary for a project to take place can be justified if the place and the project will ultimately enhance our collective ability to see through to the real, and if those engaged in and affected by the project agree to its undertaking. The key issue is of course what is meant by "collectively," "ultimately," and by "agree." These issues will be expanded upon later, but at this point we can say that occupying space, and closing the doors to our offices and laboratories, our homes and businesses, is justified if the projects that take place are intended to en-hance the view of everyone who wishes and is able to know and if there is an open system of consent and involvement. Of course, we can know if these projects are going to accomplish this only after they are com-pleted, but there are indications of the likelihood of success. One is if the project itself seems to contain the value of "seeing through" among its goals; another is if the project is not to last indefinitely; and yet another is the consensual nature of those involved in the undertaking. Again, the Iron Curtain does not meet these requirements, and most certainly not the last one.

Seeing through does not always have to be direct or immediate. Every society has ways of monitoring what takes place to determine if it meets a set of standards. In our culture, rights of privacy protect our homes from unreasonable searches and intrusions. But these rights are not absolute; they can be breached given just cause. Suspicion can arise through indirect evidence. Even though we raise our children in the privacy of our homes, schools can monitor our children's behavior to see if they are properly cared for; if they are not, then social workers and law enforcement authorities may enter our homes to see what is happening. Similarly, a scientific laboratory at a university is expected to be used for specific kinds of research, and it would not be long be-fore university officials knew if this was not the case. Private laborato-ries may not be so closely monitored, but if they produce things that contravene the laws of the community, they may be found out and shut down.

Instrumentally, social monitoring of what takes place is justified in the general sense that society has given license to and thereby has the right to somehow inspect what takes place. But this view alone does not prevent a culture from inspecting and condoning the use of place for the most heinous crimes, if they are of value to the projects of that society. Consider how the American South, and I mean here the pre–Civil War propertied class and their political representatives, valued slavery. This allowed for the existence of a vast geographical infrastructure of slave plantations that was scarcely opaque. Virtually everyone knew in general terms what took place, and Southern states enacted complex laws to monitor plantation life in order to ensure that the project of slavery was carried out. Or, consider Nazi Germany's laws that allowed the government to monitor and control what took place in homes and schools. Most Germans did not seem to object. Monitoring in these cases extended the powers of evil. But monitoring can also disclose evil.

Variety and Complexity

Monitoring what takes place, or even reducing opacity, does not then assure that what takes place is good. But without the chance of seeing through, there is no means of judging at all. Yet, the seeing through, even with our caveat, needs the corrective power of the second criterion for it not to push us too far in the direction of transparency and homogeneity. The second criterion argues that it is better to have a more varied, diverse, rich, plenitudinous, and complex reality and world, with more different and interesting places, and with a variety of points of view from which to see it. In shortened form we can call this the value of *variety and complexity*. This alone sounds attractive, but as with the first, it needs to be checked and balanced.

The value of variety and complexity works on two equally important levels. It refers to places (and projects) that we create, and to the viewpoints that they help mold. It involves then a world to see and points of view from which to see it. Consider the world as seen. Reality is magnetic because of its complexity. The mysteries of a varied and complex reality beckon us to see through, while a simple, dull, and empty world does not. We assume moreover that this variety and complexity is inexhaustible. We might even say that we are interested in the real because its variety and complexity *is* inexhaustible. Inexhaustibleness lends authority to the real and makes it compelling.

Variety and complexity must not be impervious to understanding. Parts of reality must yield to comprehension, or else the real would be something from which we would want only to escape. Understanding this variety and complexity seems to require the assumption that everything in reality is ultimately connected as a single whole. Believing this ultimate unity appears to be essential in order for us to approach the world. We become more confident that we have made progress in knowing the real when we see the complex relations among parts and whole to be generated by less complex ones. That is, when we can understand how the complexity, richness, and interconnectedness of the world at one level is produced by a simpler set of relations at another, we believe we have seen the real more clearly. Indeed this clarity reveals an underlying elegance to reality, which we strive to find and express in different ways through science, art, and religion.

Does this mean that often the best kind of contribution is a simple one, and does this not contradict the second criterion of intrinsic judgments—that of variety and complexity, especially the part on complexity? Certainly providing something simple is often the most valuable contribution, but only if it ties other things together. This makes it both useful and elegant. Great mathematical and artistic works do this. But the need to tie things together stems from the original variety and complexity of reality, and this quality of the real, which is infinite and persistent, makes this quest perpetual. We can never get to the bottom. We may progress in seeing through, and then we must see through again, and more deeply. If reality were already simple, there would be no need to discover, and if it were completely knowable, then the distinctive qualities that make us human would no longer be of use.[1]

The world's richness and complexity are not fixed and static. Nature itself changes, and we do too. At the terrestrial level, the projects we undertake require places, and in creating them we can increase the world's complexity—if we create them correctly. In this way our geographical agency in even the smallest and most intimate scale can expand the real and increase our delight in it.

Variety and complexity of place is important not only because it provides a compelling world to see, but because it helps promote multiple points of view. Having different approaches to reality is essential for us to see reality. These approaches or views moreover must be open and public and given as gifts. Seeing the world clearly can never be an isolated project. We cannot sift and winnow without conversation and a sharing of knowledge. As reasoning creatures with open minds, we use reason to evaluation views that lead to the possibility of a clearer vision that more closely corresponds to some facets of the real and the good. This clearer view would be one that is accessible to all and that reasoning creatures would assume. This process can then create the possibility that we can become less situated or less partial, but only provisionally so. If we accept this less partial view, and if the process continues, again it does not mean that we will all ultimately have only one view because now "we know it all." Reality is infinitely deep and complex, and we are constantly changing and, I hope, adding to it with our geographical activity. Hence the acceptance of a clearer view is still provisional and partial in terms of what there is yet to know. A view that we can agree is clearer than an older one still opens up only a section and layer of reality, and becomes partial relative to the possibility of an even clearer one. And we need many other such views to open up other parts of reality. There is then truth to the cliché that the more we know the more we realize there is to know. If one is worried about the emergence of a single totalizing and inflexible view, it is far more likely to arise in a closed and situated society than in an open one guided by intrinsic judgments.

Developing new forms of inquiry and expression to capture an expansive reality continues to assure the variety and complexity of contexts and places. Consider for example how one particular realm of seeing—science—has multiplied not only what is known and our sense of what there is still yet to know, but also the ways of seeing and the places that promote them. Branches of mathematics, physics, chemistry, biology, and so on, are now far more numerous and specialized, and so too are the places devoted to their pursuits; and because the facets of reality they illuminate are understood to be provisional, each part of science continues to offer ever-increasing numbers of hypotheses and views. Or take another view of the world—the aesthetic. At the same moment that forms of art are circulating around the globe and becoming ever more accessible (raising the fear of cultural homogenization), entirely new ways of expressing the aesthetic are developing; new forms of music, the visual arts, literature, dance, drama are emerging which result from the fact that we can have access to and share, accept and reject, varieties of aesthetic modes of expression. Of course, commercialism and corporate control of media are dangers to the process of free and open access to information, but I am referring here to the logical issues of the argument (that still seem to surface as a goal in spite of worldly imperfections in power and control) about how variety and complexity can increase as we strive to see more clearly.[2]

Interest in variety and complexity does not have to be driven by purely intellectual matters. This second virtue, like the first, can be of practical help. We may live better in a more varied and complex world. Biological and ecological diversity is certainly important to the quality of life. We need to have a richly diverse ecological reality to assure that we can live full lives. And the existence of different projects and cultures allows for specialization and sharing. But variety and complexity, with all of its advantages, is not a value that exists alone. Its significance must be seen in terms of its relationship to the value of seeing through, just as the value of that part of intrinsic judgment is influenced by variety and complexity. Let us examine more directly how these are dynamically related.

The Dynamics of Seeing Through and Variety and Complexity

The term "judgment" and the idea of the joint application of "variety and complexity" and "seeing through to the real" could suggest that the theory is static, detached, and final, while it is in fact dynamic and relational. The judgments are intended to animate our actions from the start. We can see that it is so by noting that the judgments are attempts to make a compelling good more accessible, and a compelling good is one that acts as a lure and sets us in motion. The problematic also stresses how we are active creatures transforming the earth, and intrinsic judgments are intended to become part of our motivations for creating places. In addition, there are innumerable ways of creating places to meet these criteria, so the judgments encourage innovation, or variety and complexity, that in turn provide greater awareness and more ways of seeing that create yet greater variety, and so on. Thus intrinsic judgments provide a direction and help propel us along it.

As the previous section suggests, the core of the dynamics lies in the tension between the two criteria. How this is so needs to be put explicitly by reconsidering some of these examples. Consider first how variety and complexity is regulated by and justified through seeing through. A varied and complex reality encourages us to see the real by holding our attention. A sharing of varied and complex views, resulting from engagement in varied and complex projects and cultures, helps us see more clearly. Yet, this can be accomplished only if the places that create this complexity and diversity do not overly obscure our vision and prevent us from understanding and communicating what we create and see. Balancing "variety and complexity" with "seeing through to the real" establishes the right proportions of opacity and transparency (or, thick and thin, if one stresses the opacity and transparency of the weave, rather than the boundaries). Even though projects need to be bounded, a complex and diverse world is ultimately a connected one where the boundaries of its places are neither too transparent as to discourage projects, nor too opaque as to disconnect them from the wider world and prevent us from seeing through. Diversity is of value, then, not by itself, but as part of the world and as a means of expanding everyone's awareness, and this process of being part of a whole that expands awareness also sets limits on diversity. A cultural enclave that produces superstition and narrow-mindedness, that disparages strangers and isolates its members from the rest of the world, may be contributing to diversity, but at the expense of narrowing vision and disconnecting parts. A slum or an opium den can be "different" and "exotic" to those on the outside, but it imprisons those within. In a similar vein, the differences created by the Soviet experiment were more than offset by the obstacles it presented for those inside and outside to see through to the real.

On the other side, seeing through is tempered by variety and complexity. If places were so transparent, if we could see through without any effort, then significant projects could not be undertaken. Everything would be constantly open to public scrutiny, or to put it another way, we would constantly be "on stage." There would be virtually no "backstage" in which to rehearse, no place to withdraw and think, no place to experiment. Complete transparency then would create a dull and monotonous world, one which lacked variety and complexity, and which would finally not encourage us to see through, for there would be little of interest to see, and would be the only point of view from which to see it. An overly transparent world, paradoxically, narrows our vision and contracts our horizon. Another way of putting this is that the joint application balances the relationship between being more and less situated. The qualities of place that promote situatedness are good and necessary for the creation of varied projects and viewpoints, but so too are the qualities of place that allow us to be open, receptive, and more objective.

As the joint application of seeing through to the real and variety and complexity help in balancing transparency and opacity, it also helps navigate between absolutism and relativism. On the one hand, the variety and complexity criterion encourages differences in viewpoints, but the seeing through requirement discourages relativism among them. And, on the other, the provision of multiple perspectives encouraged by variety and complexity diminishes the chances that the seeing through criterion will lead to the absolutistic and oppressive claim of one view seeing absolutely correctly; it will promote instead the idea that all views are to be taken critically and provisionally.

The joint application of the two criteria thus provides a guide or regulative principle to the type of places we should have. As a regulative principle, it performs a similar function to Kant's Categorical Imperative. Kant proposes the Categorical Imperative as a basic criterion that our reasons for doing things must meet for them to be morally sound reasons. The Categorical Imperative does not say what is good, but rather what test our reasons or motivations must meet for them to be good. In Kant's terms a reason for undertaking a project is a good reason if it can be universalized—or willed as a universal "law"—that everyone would agree to follow. That is the essence of the Categorical Imperative.[3] This would then be applied to the reasons we use to justify our actions and would thereby determine if those reasons are themselves good or moral ones. For example, suppose that in order to borrow money, someone makes a promise knowing full-well it cannot be kept. Would this type of behavior stand up if it were universalized in the sense that it becomes advocated in all cases like this and for everyone? Kant would argue that it wouldn't because not keeping promises destroys the very possibility of trust and respect for others. If a maxim or rule passes the Categorical Imperative, then it is a good. Otherwise it is not.

The intrinsic geographic judgments jointly act as a geographic categorical imperative. They ask if a place or system of places is providing both the qualities of seeing through to the real and variety and complexity. They do not themselves specify what exactly seeing through and variety and complexity look like on the ground in each particular circumstance. They cannot and should not. Potentially countless types of place could be good, and the balance for each place can differ. Yet they do provide a means of assessing a particular instance to tell us if it seriously violates these principles, and they also point in a direction that could help in altering and creating new places. Moreover, applying these criteria jointly orients the particular virtues within a place. It makes the various uses of truth, justice, and the natural more complementary and, as we shall see, oriented toward altruistic qualities.

In providing this judgment and guide, intrinsic judgments are not acting remotely and statically, for they are used not only as a means of judging but as a means of animating and directing our place-making efforts. The joint application contributes to new places, projects, and ideas that eventually expand reality and our awareness of it, and which in turn lead to more varied and complex creations. The different points of view these provide can stimulate the imagination to arrive at a still more comprehensive view that includes the differences, indeed making them even more interesting by expanding their resonances, and this more comprehensive view is only a stage, for it in turn must be examined from and compared with yet other points of view. The process is continuous.

Consider, for example, how this can happen in the case of organic coffee production. The development of programs certifying and promoting organic coffee attempts to satisfy several goals. It promotes a criterion of justice that assures that individuals and families who

produce the coffee receive the revenues; they encourage local autonomy and decision making; they promote ecologically sound practices
of coffee growing; and they promote an organic product that may be
healthier to the consumer. These goals require that the process be
made as transparent as possible. Those who help do so include local
growers who must examine what takes place in neighboring fields, as
well as regional and international agents who make sure the practices
conform to general standards, and assure that the public who consume the coffee know something of the local practices. The important
point is that this openness and heightened awareness does not lead to
homogenization. Rather it actually opens up the possibility for difference and diversity. The goal of promoting sound ecological practices,
for example, requires that these be virtually redefined for particular
contexts, adjusting them to properties of soil, slope, and various in-
dices of biodiversity. An awareness of the agricultural practices in a
neighboring farm can be used creatively and differently in yet another.
The local farmers, in turn, can become aware not only of other agricultural practices, but also of the tastes and styles of consumption of
the distant international coffee-consuming public. This geographically
dispersed and varied public, as well as the local coffee growers, can
then use this knowledge of production and consumption as a means
of expanding their awareness and appreciation of ways of life, of creatively changing practices, and of being better moral agents because of
their heightened awareness of the consequences of their actions. Of
course for this to happen, they must make an effort to use intrinsic
judgments not only as a means of evaluating practices, but of inspiring
their place-creating activities each step of the way.[4]

This dynamic and expansive quality of intrinsic judgments can be
seen in an economic practice, such as coffee production, and in the less
tangible and more "symbolic" systems of language itself. The meaning
of words, even for feelings that we all may share, draws on our local
practices, and the circulation of these meanings deepens our awareness
and appreciation of both these virtually universal emotions, and of different local practices. Consider, for example, the feeling of maternal
love.[5] Different contexts and symbolic systems allow us to take this general emotion and refine and differentiate it. The Maasai, a cattle-herd-
ing people of East Africa, address a child fondly as "my umbilical cord."
Americans often fondly call their child "honey." The result of a mutual
awareness of these different linguistic usages (that draw from different
contexts) is to make us more deeply aware of motherly love and its
meanings. It can inspire yet another metaphor for this emotion, and
lead to appreciation of the contexts from which these expressions for

love were drawn. But again, this can happen only if we are open to this variety and complexity and if it allows us to gain insight and see more clearly. When guided by intrinsic judgments, places can then help us see through, share, and open us to new possibilities.

Evil Places

Implicit in our discussion is that once the theory points to the good, we would move in that direction. But a theory of the good cannot force; it can only persuade, and its persuasiveness can be enhanced if it can also identify the wrong direction and describe why it is not appealing. Making an evil place is a process involving continuous contraction of awareness and variety and complexity.[6] We have already observed how this process works (and will return to these cases in the next chapter). Here I want only to abstract and categorize the end point and show how these types of places are related to the theory.

As we stated, there are three kinds of evil places. One results from the absence of "seeing through," one from the absence of "variety and complexity," and one from the absence of both. Though most systems of evil exhibit more than one of these, it is important to take the first two separately.

The first one follows from the opposite of "seeing through to the real" and refers to places that narrow and obscure our vision. Often it is important to distinguish between using place to narrow the vision of others, the victims, and using place to cut off one's own vision, as often the victimizer also becomes the victim to his own narrowing view. We have seen how those in charge of the Iron Curtain narrowed the view of those both inside and outside, and also their own, as they too had access to less information. This was also the case with Nazi Germany which not only isolated and contracted the world of its victims, but also contracted the world of its own citizens through censorship and the role of terror. The same is true in the case of American slavery, where the plantation narrowed the world of the slave and its rules of censorship narrowed the world of the planter. To these can be added China's Bamboo Curtain and North Korea's isolation. In its extreme geographic form this narrowness and obfuscation can be found in moves toward *autarky* and *secrecy*—intentionally cutting one's society or a portion of one's society off from the rest of the world. Again, narrowing one's vision is not the only evil quality embodied in these complex places, but it certainly is important. And what of smaller-scale examples? We can cite, for example, the cases of cults that seduce young children to join and then prevent them from contacting those on the outside, or the case of a home that instills in the children a hatred of others that narrows their scope of the world.

The second one follows from the opposite of valuing an ever more varied and complex geographic reality and occurs when the variety and complexity of places are diminished. The most distinctive geographic form this takes is a *tyranny* of one type of place over other types.[7] This is found again in the examples of Chapter 4 where virtually every place within these regimes had its mix of virtues twisted by the goals of the entire system. In the Nazi case, homes, schools, and workplaces became alike as instruments fulfilling the goals of the party. A subtler variant of this evil occurs when a model of one type of place becomes the dominant metaphor for very different types of place and, as this metaphor (and other qualities of language) becomes adopted, the places using them tend to become alike. The result is a loss of various views about reality and combinations of virtues. An example is the business model in contemporary society that is being urged upon schools and hospitals so that students and patients are turned into customers, while teachers, professors, doctors, and nurses are turned into producers and providers of products and services. If this imposition of one form of language on another is successful, then schools and hospitals can be expected to become more like places of business, and so the differences among them diminishes. Another example occurs when the products of a place tend to homogenize facets of the world as many would argue is happening with certain forms of mass production and advertising.

The third geographic evil follows when both geographic criteria of the good are violated. The geographic form this violation takes is a general and constant *transgression and chaos*. Certainly change for the better is accomplished by some transgressions, but so too is change for the worse, and if there is only transgression, then no projects can take place. We will not be able to see through for there will be neither a viewpoint nor anything to see, and an inchoate nothing is not the same as complexity and difference. While transgression and chaos were qualities experienced by the victims of the totalitarian tyrannies of Nazi Germany and the Soviet Union, and also of slavery, it is perhaps most intense in countries that experience collapse, such as the case of Sierra Leone, and in countries under the threat of terrorism. By making anyone and everyone its victim, and instilling a fear that any place can be violently transgressed, terrorism attempts to create chaos.

Being Human and Moral

If we consider again the examples of evil places, one might ask: Why do we need to stress the criteria of awareness and variety and complexity when these clearly bad places can be condemned directly? We can say that a concentration camp is a place that encourages murder and torture, and a slave plantation is horrendous because it diminishes human dignity and freedom, and both are detestable because they increase human suffering. Why not have the pursuit of pleasure, the feeling of compassion, or some other emotion serve as a guide for moral judgment rather than using "seeing through" and "variety and complexity"? Why can't these be the core of being good, and their opposites, the core of being bad? These are certainly more familiar categories, so why aren't they the foundations of the theory?

My response is that theory is not based on them because they do not go to the heart of moral agency. We do not consider animals to be moral agents, yet they seem to seek pleasure and avoid pain and satisfy emotions and passions. We do not hold them to be morally responsible because they are driven by instinct and situations. They could not really do otherwise. Pain and suffering in humans moreover are not necessarily immoral. Some of it is self-induced. We may put ourselves through the worst hardships in order to do what is best. And, alleviating pain and suffering is not in itself a moral act. An evil dictator can provide us with many of our needs, keep us happy, fed, and protected, provided that we then do his bidding. The central point again is that to be a moral agent means we must have awareness and free will so that we are responsible for our actions. It is this capacity to be aware and to choose that is necessary to do good. Once we recognize this crucial point, we are then able to morally assess pain and suffering, or the pursuit of pleasure and happiness. Pain and suffering are bad things when they diminish awareness and the capacity to choose. Diminished awareness also explains why we harm others. We cause suffering, kill others, treat them unjustly, humiliate them and take away their dignity, and degrade nature because we do not fully understand what we are doing. Yes, we are gripped by vices such as greed, envy, hate, lust, and sloth and by institutions that have their own defects, but the tightness of the vices is due to our ignorance; awareness of our own defects lessons their hold on us. What other ground provides a justification for moral critique and a hope for moral improvement?

True, doing good draws on sentiments and especially on the sense of compassion. But these emotions alone cannot serve as a guide to action; in fact they can lead to serious mistakes, as when maternal love and affection may prevent a mother from turning her offspring into the police for being a drug dealer and murderer. Sentiments must be weighed and guided by a more abstract sense of the good, though when one does the right thing the act is accompanied by some sense of peace and satisfaction that outweighs the pain that doing it might cause.

Reinforcing the importance of awareness and choice as the central components of the moral theory is that these make sense of other particularly human qualities that seem to not have a clear evolutionary (and thus instrumental) function. I am speaking here of our ability to imagine, to represent symbolically, and to exercise reason. Other species avoid pain, experience pleasure, and survive quite well without the capacity to think abstractly and represent the world. In fact, thinking about the world, imagining new worlds, and reasoning and making choices on the basis of our imaginations can be our greatest source of affliction. Our mental capacities and imaginations seem to have far outstripped their evolutionary function. My point is that existence of, and interrelationships among, these qualities make sense when they are seen as providing necessary support for being moral agents. (In functional terms, being moral provides them with a function and purpose.) For example, reasoning in the human case is the capacity not only to make decisions, but also to hold the alternatives in one's mind, including alternatives that may not yet exist, and to be conscious of and reflect about them. We not only decide, but we also come up with reasons for our decisions, and this again means we are conscious, reflecting creatures who can represent to ourselves and others what is and what we think can or ought to take place. To hold these things in mind and to communicate them, even to ourselves, requires the capacity to symbolize through a variety of forms, including ordinary words and language, and specialized symbols and languages such as science and art. Using symbols, representing the world and imagining alternative ones, and giving reasons for our decisions are necessary for reasoning. That we do all of this, and especially give reasons to justify our decisions, makes us distinct, and makes sense if we are moral beings.

Having these capacities allows humans to do good and possess (or instantiate) goodness. And if other creatures posses these capacities then they too would have to be considered moral agents. They would possess the dignity and worth of such agents, and also bear the responsibility of being morally evaluated. There may well be other beings in the universe with these capacities. The value and dignity of their lives would then be as great as ours, for these derive from the capacity to be moral.

In the absence of the capacity to reason, have a will, and make choices, our actions would simply be prereflective responses (conditioned by such things as drives or instincts) to forces that we do not initiate. Our actions then would not be due to us as conscious and free beings; rather we would be caused by these forces.

The components stressed in the moral theory reinforce the value of human life—of treating human beings as ends. This applies to all humans, and not only the ones nearest me, the ones I know, or to whom I am related. And it is not only those who are alive now, but future generations as well. Those who will follow will also have this moral worth, and we have the same moral responsibility not to diminish their awareness as we do to the present generation. So the theory and its intrinsic judgment provide grounds for being as expansive in our care and concern for others as possible.

There is yet a further benefit that comes from the theory. Its core ideas move us beyond a moral concern that focuses entirely on justice to a focus also on the idea of truth and properties of nature. Justice is undeniably a central issue in morality and it has been the preoccupation of most contemporary moral theorists. But it is only one of three moral realms that geography brings to our attention. The other two are the search for truth and our relations with the natural. All three are part of what is real and good, and neither of the two is necessarily subservient to justice. A heightened awareness of the real and the good is the means by which we can emphasize truth and the natural along with justice, and also the means by which these virtues become clarified and seen not as relative, but as contextual.

Each of these issues supports the theory's emphasis on awareness, and choice. Let us now take a closer look at how these are related to other parts of the theory—namely, the attraction and ineffability of the good.[8] If the good is compelling, then in what way are we choosing, and how can we not be good?

REASON, CHOICE, AND THE ATTRACTION OF THE GOOD

How do we reconcile free will and a compelling good? This section restates this problem by drawing on two extremely important images— one from the Biblical story of the Garden of Eden, which stresses the importance of choice; and the other from Plato's Allegory of the Cave, which stresses the compelling quality of the good. Both of these stories are complex and even contradictory; here I wish to single out only those themes that have a bearing on the argument.

The Garden of Eden story is strong on choice, but weak on why we would or should choose good. This is because it is a religious story: we simply must accept on faith that God's often-inscrutable acts are good. In the Garden of Eden, we find Adam, Eve, the serpent, and, most importantly, the fruit that grows on the tree of knowledge of good and evil. This tree and its fruit are an allegory for at least two qualities of human awareness. It represents an arrogance that drives us to think we can really know everything and become godlike, which will doom us to failure; and it also represents the need for humans nevertheless to strive to become aware so that they can make a choice. I will focus on the second. If Adam and Eve did not eat of this fruit, they would have been obedient to God, but blindly so. Blind obedience may be desirable in some religious circles, but it does not seem to have been so in the Garden, for God provided a choice. Blind obedience does not exercise the critical faculties to discern what is good and bad; it merely compels one to follow a commandment. Giving Adam and Eve the choice meant that God did not think that good can come from obedience in the absence of awareness. He planted the tree and the serpent and gave Adam and Eve free will. If they were to have followed his commandment, they would have done so without being aware of why disobedience was evil (just as a pet may obey our command without consciously doing good). To do what one is commanded to do when one does not know what is good (or evil), is not really to be good, but to be obedient (again, like a pet). The choice they made of eating the fruit was disobedient; but only after it was made and they became aware of good and evil could they see it was evil, and then only if one believes that God's commandments, though inscrutable, are good. One can say then, that God gave them the opportunity to disobey so that they would know and be responsible for doing what is right, and to realize that listening to God would help them know the difference. Our original sin is the fallibility of our judgment.

The choice made Adam and Eve move from the status of pets to that of humans. They became aware of good and evil, and also of their own vulnerability in selecting. They now knew they were naked. This capacity to choose, or this capacity to be aware and to imagine things that may not yet exist, has risks. We may imagine wrongly, and choose incorrectly. And once we imagine and choose, we then create an environment that is full of possibilities and also of mistakes. We are no longer in the perfect stable world of the Garden, but in the real world, striving to make it conform to our ideals.

This Biblical story establishes our problem of choice and awareness, but it does not make a strong case for the ineffability of the good and its magnetic quality. Rather it relies on trusting the word of God. For the attractive qualities of the good, we should turn to the Allegory of the Cave in Plato's *Republic*. This is a broad philosophical work whose focus is to create a state and society—a republic—that is informed by the good. It assumes humans are reasoning creatures, and focuses attention on how it is that the real and the good attract, not by the command of a god, but by the good being singularly clear and compelling to an open and reasoning human mind. Religious arguments follow god, and philosophical ones follow reason and the good. To explain the magnetic qualities of the real and the good, Plato would have us imagine that human beings are inside a cave, chained to their places near its deepest end, and staring at the cave wall. Behind them burns a fire, and between them and the fire is an elevated walkway along which move people carrying all sorts of objects. As they do so, they block parts of the light from the fire and cast a shadow onto the wall. It is this shadow, and the shadow cast by the chained prisoners themselves, that these people see. This is all they know of the world: the shadows on the wall that are in fact cast by themselves and the objects and people behind them and in front of the fire. They take this vague, dim, shadowy world to be reality.

But one among them is unchained and forced to turn around and face the fire. At first, its light is too blinding, and he sees nothing. Then he discerns the objects on the stage, and finally the fire itself. He is then able to understand for the first time that the shadows are not real, but only paler forms of reality. Seeing these objects helps clarify his world, but only a little, for he is soon moved further through the tunnel, and as he does so he gains an understanding that these objects and the fire are within a cave, and finally he emerges into the light of day. This daylight, like the fire in the cave, is at first blinding, and for a while he is comfortable seeing only reflected light, the light sparkling on the water, the moon, and the starlight. As his eyes become adjusted, he sees that this outside world is far more vivid and its elements clearer than when he saw the objects in the cave and not their shadows illuminated by the fire. Outside, it is the sun that illuminates, and as soon as his eyes have become adjusted to its brightness, he notices this ball of fire in the sky; yet (although Plato does not say it, we must assume), it is too bright for anyone to see directly, completely, and continuously. Rather he sees only facets of it, a ray here and there. Still it is enough for him to feel its pull. He is drawn to it, and the world it illuminates.

The sun stands most distinctly for the real, and the other images in the allegory for levels of awareness of reality, but, as Plato states emphatically, it also stands for the good—the two being inextricably intertwined. Indeed the good is the source of illumination for the real. As Plato himself remarks, "knowledge of the idea of good appears last of all, and is seen only with an effort; and, when seen, is also inferred to be the universal author of all things beautiful and right."[9] The allegory provides a classic image of the ineffability and attractiveness of the real and the good. They can never be seen in their entirety; like the sun, they are too brilliant and they stand as a source, a guide, and as an object of wonder. This person could have remained above ground to enjoy the light of day but Plato has us imagine him returning to the cave, to help those who are still in the world of shadows. Returning, the individual will again suffer temporary blindness and disorientation. Instead of his eyes adjusting to the brightness, they now have to adjust to the dimness of shadows. Even after adjustment, a person who returns will not see the same things in the same way, and cannot help but assist others to see more clearly. Imparting this gift makes everything else gain in brilliance. That is, the real and the good reach their full powers only when made public and accessible to others.

Though in the allegory the individual is forced to turn around and face the good (instead of depending on his own will to do so, as the theory requires) and the good is seen from only one person's view (instead of having a variety of views, as the theory requires), it stands as the classical description of the compelling and ineffable quality of the real and the good. In other writings, Plato puts the same point more directly. In Protagoras (352b) he asserts that "knowledge is something noble and able to govern man, and that whoever learns what is good and what is bad will never be swayed by anything to act otherwise than as knowledge bids . . ." and in (358d) he goes on to say that "then surely . . . no one willingly goes after evil or what he thinks is evil; it is not in human nature . . . to do so—to wish to go after what one thinks to be evil in preference to the good . . ."[10] People who do wrong simply haven't seen clearly enough. They have somehow been deluded. Kant too, though in a different way, believes that the ineffable good or moral compels us once we were led to its necessity by pure reason. Choosing is where the moral enters as a force, and only if we will it to. Otherwise, without this will to reason, we are driven entirely by a web of causes that make us their victim. G. E. Moore argues that not only is the good compelling, it is "a real constituent part of the world."[11] In geography, it is Tuan who has been the strongest advocate of a good that is real, ineffable, and compelling. As he puts it: "Its very lack of specifiable content imparts to it the authority of the impersonal and the objective. A vision of the ineffable good . . . provides a safeguard against intolerance and moral stasis. What we need is a powerful lure which limits the indecisiveness of freedom and yet does not enslave or blind."[12]

The assumption of the infinity and ineffability of the good makes it undogmatic. It does not provide commandments written in stone, for those become more like a yoke around our necks than a means of liberation. The good must not chain us, but "set us free." Its power is not to predetermine what we do, but rather provide a lure that pulls us, or a magnet that orients us, in the right direction. If there were no such lure, no such magnetic force that was real and that existed apart from particular projects, then we would be back to instrumental judgments and to the conclusion that morality is relative, a product of power, and nothing more than mores or customs. And, if the good, like the real, acts as such a lure, then it is compelling. This is why evil is understandable as a lack of awareness of the real and the good. This is the only conception of evil that offers society a means of correcting mistakes and making progress. It assumes that through attending to and educating those who have done wrong, we can help them become better. It also must assume that increasing awareness requires a variety of places that can generate different points of view so that what is seen can be sifted and winnowed. Seeing clearly depends on variety and complexity. While this is the case, we still have not come to grips with the basic problem of why we choose evil.

There are no ready solutions. I believe, though, as I have already suggested, that if there is one at all, it begins with clarifying the problem of self-deception as the source of evil. The importance of self-deception follows from our previous points. The theory argues that without exception, *we do not choose evil willingly*. The distinction between good and evil depends on it. If we find evil as attractive as the good, and so do evil willingly, then I see no way of telling the two apart: one could be substituted for the other. And, just as Milton's Satan was involved in self-deception, so too, I will argue, was Hitler and Stalin. They did not see the real and the good clearly. It is impossible to prove that this was the case, but the evidence can be interpreted that way, and to do so provides greater breadth and consistency in explaining the moral import of our actions. This is evident as soon as we see that a weaker assumption—which challenges the idea that the good is always compelling and expects that sometimes some of us choose evil willingly—makes the difference between good and evil relative. A weaker assumption also makes it impossible to rehabilitate the evil-doer through instruction on what was right and wrong, and so challenges the role of awareness and education as correctives to wrongdoing. It also characterizes those who do evil willingly differently from the rest of us. Unlike the rest of us, they must be moral monsters. (I am not of course saying that there are no people who in some essential way are "hard-wired" to inflict pain and suffering—but such people are not doing it out of choice. Their "wiring"—or disease or insanity—does not allow them to do otherwise. Hence they cannot be blamed for what they do. They simply are not responsible. Though we must remove them from society, they are not choosing to do evil. By most accounts, Hitler and Stalin are not that sort. They did have the capacity to do otherwise. My point is that since no one could do evil willingly, they were involved in massive self-deception).[13]

For these reasons, the theory retains the strong claim that the good is compelling and that we do not do evil willingly. While this removes the problem of choosing evil over good, it still leaves the problem of explaining why it is the case that often we do not choose good. This is where self-deception plays its part.

SELF-DECEPTION

The theory shifts the site of free will and responsibility from choosing evil willingly, which it denies can be done, to willfully not trying hard enough to be aware and open-minded. While not trying hard enough to be aware, or even deciding not to be aware, is not in itself choosing evil, it does make our future selections more likely to be evil. It is at the point of deciding not to be aware that we are held morally responsible, and my argument will be that self-deception is the principal mechanism by which we diminish awareness.

I cannot offer a definitive explanation of self-deception, but rather a provisional description of it that may clarify our problem of good and evil. I believe that for self-deception to occur there must be, as Jon Elster put it, "some breakdown of internal communication in the mind"[14] and this breakdown is part of a process of mental compartmentalization. An internal mental barrier is erected between parts of the mind separating competing arguments or interests. These moreover tend to be asymmetrical in their claims. That is, if the two or more parts were joined, we would have to conclude that only one part is right. But by mentally separating them, we can mislead ourselves and believe that another part is right. This may happen because the part that is not right is attractive to us for other and instrumental reasons; it fits into some other schema that we greatly value, and so we do not want to make the effort to confront the facts.

This idea can be stated more precisely this way: self-deception occurs when a person holds two (or more) contradictory types of evidence in mind—one set supports "p" and the other its opposite, "not-p." Even though the evidence for "not-p" is far stronger than the evidence for "p," the person will keep her knowledge of "not-p" compartmentalized and apart from her knowledge of "p," and even avoid calling it up. This can be called a weakness of will.[15] It is a useful way of describing self-deception, and as a description (but not an explanation) of the condition it is familiar and accessible, as the following examples show.

Suppose a parent loves her grown-up child very much. This child is kind to her and the other members of the family. But suppose that the police offer her overwhelming evidence that her child is a drug dealer and cold-blooded killer. She not only loves her child, but wants to believe her child is good, so she takes the evidence of her child's kindness to her as supporting this belief—this "p." She also does not either want to know or think about the evidence from the police—the evidence of "not-p." She keeps the two apart, and even denies the significance of "not-p" in order to believe "p." In doing so, she is exhibiting a weakness of will and is involved in self-deception. This is self-deception because her desire not to be aware and confront what is there enables her to hold on to a belief that she knows she could not otherwise justify. She is therefore mentally compartmentalizing her world in order to not see clearly.

A crucial point (that is at the basis of the next chapter) is that self-deception is linked to geographical compartmentalization. In deceiving herself about her child, she is *not* mentally confronting the evidence, and she is also not physically *placing* the evidence before herself and her child; conversely, by keeping the evidence from view, she does not have to confront it and forsake her belief in "p." Self-deception does not work without geographical compartmentalization and keeping things from view. Self-deception requires that the deceived not see a landscape with varied points of view, and diminishing the actual variety and complexity increases the possibility of self-deception. There are related roles that geography plays in this process. Using place to create an appearance that keeps the truth from view is another.

Suppose I am a slaveholder. Even though I may have a gnawing suspicion that slaves are fully human beings, or "not-p," I accept the white slaveholder's view that they are subhuman, or "p." I do so by ignoring the meaning of the overwhelming evidence that supports "not-p,"-evidence that includes rules the slaveholders enact and that slaves must obey and are able to obey only if they are in fact fully reasoning creatures (or else these rules would not be necessary): rules such as not allowing slaves to learn to read. I avoid this evidence because doing so is useful to my interests. I enjoy the life of a slaveholder, recognize it as a profitable economic system, and to admit that slaves are human and reasoning beings would make me see that the only foundation for slavery was a lie that allows an extremely cruel form of exploitation. Could I stand this? My point is that this recognition would lead me at least to want to renounce slavery in principle. I agree that it may not lead me to emancipate my own slaves, for that might be an enormous financial and social sacrifice, and might not even be possible in some slave systems. But it would lead me to push toward a resolution in favor of emancipation. To avoid this move in the right direction, it seems to me that I would have to hold onto the false claim that slaves are inferior and thus deny that slavery really is evil. This I believe is what happened in most cases in the American South. While supporters of slavery recognized that it was a profitable institution, they also thought it was morally justified because of the inferiority of the slaves. The 1861 statements of the president and vice president of the Confederacy make this point explicit. President Jefferson Davis argued the economic side, claiming that Lincoln's proposals to exclude slavery in the new territories would make "property in slaves so insecure as to be comparatively worthless, . . . thereby annihilating in effect property with thousands of millions of dollars." Alexander H. Stephens, the vice president, argued the "moral" side, claiming that the foundations of the Confederacy do not rest on the incorrect assumption that all men are created equal, but rather "founded upon exactly the opposite idea; its foundations are laid, its cornerstone rests, upon the great truth that the negro is not equal to the white man; that slavery, subordination to the superior race, is his natural and moral condition. This, our new government, is the first, in the history of the world, based on this great physical, philosophical, and moral truth."[16]

A self-deceived group will often believe that they are actually seeing clearly. How though can we distinguish between someone or group that is self-deceived and believes they are doing the right thing, and really doing the right thing? Or to put this in terms of an instrumental versus an intrinsic distinction, how do we know when an instrumental position is mistakenly taken to be an intrinsic one? As we noted before, the core of the answer lies in the degree to which those holding a position subject it to the criteria of intrinsic judgments, but before I go on, I must point out first that not all mistaken ideas of the good are instrumental. It is possible simply to be mistaken. We may not be seeing clearly, or clearly enough, and intrinsic judgments would make us aware of the fact that this is likely for any and all of our beliefs. But it is when a belief can be shown to be self-serving that it is particularly suspicious. The coincidence of a "moral" position with self-interest is not proof that the moral view is instrumental. It does though strongly suggest so and demand that the view be further justified. And if the holders of that view are well-intentioned, they then would be open and willing to listen to other views—a possibility that comes from the existence of variety and complexity (and not employ self- or group censorship)—and, in the course of sifting and winnowing, be willing to change their minds even if it means that a new position they may come to hold is no longer instrumental to the original project. That is, they would make every effort to avoid self-deception. This of course is not a foolproof test, but it is the way the theory approaches the problem.

Self-deception is not only of theoretical importance in explaining how the good can be compelling and why we still may not follow it, but it is also a part of the logic that undergirds a society's understanding of responsibility. Liberal social systems assume something like this mechanism of self-deception when they stress the value of freedom of speech and expression, and the redemptive role of education. On the one hand, they are claiming that greater awareness of the world leads to moral improvement. By the same token, it means that those who have erred and committed crimes have done so through a lack of awareness and can be rehabilitated through some form of education. Such societies often provide educational opportunities to prison inmates (or rather to those in "correctional institutions") and stress that education, rather than punishment, is more effective in dealing with those who behave badly. (What I have just said is not intended to equate laws with morality. Many laws are in fact immoral, and there are often good moral reasons to disobey them. But let us put aside this extremely important issue for now, and assume that laws are motivated by moral concerns.) At the same time, they are holding the individual responsible for making an effort to be aware: they see the individual as exercising free will. This is especially acute in criminal law. Courts hold the individual responsible unless he or she is somehow mentally impaired (as in not having sufficient intelligence or being criminally insane) or did not have the choice of doing otherwise (as when someone kills in self-defense, or when a pedestrian is killed when a car's brakes fail).

The assumption that self-deception is the point on which to focus allows society to hold both the idea that the good is compelling and that the criminal is responsible for not having seen the good (where again I am, for the sake of argument, equating illegality with immorality). The criminal knows right from wrong but did not make the effort to become aware of what was right in this instance. And the process is usually spread out as a chain of decisions that led the criminal into an ever-narrower perspective. Which one is critical may not be possible to decide, but still the assumption is that the person could make a decision along the way that would have altered the course of action. There are undoubtedly social and institutional factors that narrow an individual's context and range of awareness to the point where the opportunity for choice vanishes. In these cases, the individual cannot be held responsible for his or her actions. But when there is room for choice, the critical point is that one must choose to be aware.

Evil is far less remote when we think of it arising from not choosing to be aware (as opposed to being fully aware and then choosing evil over good). We can all understand how the incremental decisions to avoid being open-minded can lead to the morally wrong choices. This makes evil more ordinary, or, as Hannah Arendt calls it, banal.[17] Seeing evil as banal avoids setting it apart as something extrahuman; rather there are evil acts to some degree in all of us as we fall prey to self-deception. It also allows for the possibility that we can improve ourselves and the lives of others by encouraging the creation of places that help us expand our horizons. And, for the purposes of the theory, it helps make it understandable how unlike the empirical causes that force us to do things, a compelling good does not force us, for it still requires effort and awareness on our part to be drawn by it. A good that is compelling then does not contradict the role of free will.

ALTRUISM

The theory's components of intrinsic judgments, free will, and a good that is compelling are closely related to altruism. This is evident as soon as we recognize that awareness of the good cannot be a private or secret awareness. Awareness must be made public and shared. Imparting awareness, giving it away, is how the checks and balances of intrinsic judgments work. It is how we avoid self-deception. Intrinsic geographic judgments not only encourage us to create places that expand our view of reality and that increase variety and complexity of reality, but encourage us to impart these expansions as gifts. That is, intrinsic geographic judgments encourage a disposition toward altruism and altruistic gift-giving.

Pure altruism is the giving of a gift without any expectation of return. In other words it is an act that is not motivated by self-interest or by the expectation of reciprocity, as is the case with the less altruistic concept of "reciprocal altruism" and the related notion of "enlightened self-interest." But again, a note of caution. Pure altruism, as we noted in Chapter 1, is impossible to achieve (for, in the extreme, constant giving without restraint can destroy the giver, and giving without some anticipation of its use can lead to thoughtless and harmful gifts that damage the recipients), but is an ideal worth striving for, and marks the direction in which intrinsic judgments encourage us to go.[18]

Internalizing intrinsic judgments shifts the perspective of people undertaking projects within place from one of a narrower vision of instrumental use and self-interest to one that considers the place and its projects also as a gift. These will still have instrumental value, but one that is shaped by intrinsic judgments. The place itself becomes valued as a gift, the products of the places are seen as gifts, and these gifts will be part of the realms of social relations, meaning, or nature; that is, they will involve the circulation of truth, justice, and the natural.

Good places are ones that help make altruism, selflessness, and altruistic gift-giving more likely. Realizing that such acts require place also makes us aware that attributes of the gifts, while containing all three virtues, will likely stress one: a gift can be mostly part of the realm of social relations (which is what people normally expect altruism to be when viewed nongeographically) and circulate in particular forms of justice and care for others; or it can be mostly part of the realm of meaning where it is imparted as a gift of knowledge and truth; or it can be mostly in the realm of nature in which it becomes a gift of nature or natural endowments. The fact that altruism occurs in each of these realms is evident when we think about the selflessness of giving, and even the risks attached. That is, giving not only is an effort, but giving in each realm can produce heroes or martyrs.

The most obvious examples of gift-giving in terms of the social or of justice are in the form of charity and self-sacrifice. Charity can, of course, bring the giver rewards such as recognition, prestige, and gratitude, but anonymous giving, and especially anonymous giving to distant strangers, provides many examples that could approach pure altruism. Such donors would not be known, and would not expect anything in return. True, giving may make donors feel good, but if we harp on this, we then are straining the facts in order to explain away what is essentially an altruistic act. But we can go farther still into the realm of altruism. Instead of giving money or material goods, people can give of themselves to help the needy. They can volunteer their services and provide skilled care as in the case of Doctors Without Borders and the International Red Cross. A great gift is certainly the dedication of one's life to help others. One can even martyr oneself for the cause. A Mother Teresa is not only offering her services, but also sacrificing her well-being and personal safety to help the needy. Extraordinary cases of self-sacrifice are found in the heroic risks that some undertake to keep strangers from harm. Examples can be found among antebellum whites who risked their lives to free slaves, and among Christians who risked their lives to help Jews in Nazi-occupied territories. People such as these, and the less dramatic but indispensable charitable agencies and individuals around the world, give the gifts of care to rectify injustices.

People can be martyrs for justice; they can also be martyrs for truth. Socrates is perhaps the first historic figure who gave his life for free intellectual inquiry. He may be the first intellectual martyr. He did not profess to know anything—indeed all he knew was that he did not know (and by implication others did not either)—but he demanded the right to question openly and publicly, and took his own life rather than accede to Athenian wishes to have him foreswear questioning. To his name can be added the long list of scientists, intellectuals, and artists who risked ostracism, persecution, and even death to publish, or perform what they thought were insights about reality. Galileo was threatened with excommunication and imprisonment because of his beliefs about the nature of the solar system, and Václav Havel, later to be elected Czech president, was jailed under Communism for writing plays critical of the regime. These and many others believed that increasing our awareness of what the world is really like was a gift worth giving even though it put them in personal peril. They believed they were making a contribution to the public good.

What of the realm of nature? We can find generosity and altruism here in two senses. First one can find examples of those who find nature morally inspiring and want to extend rights to nature (which of course means giving away some of our own rights as a gift), and who dedicate their lives to saving or preserving species and places of nature and even risk their lives for the sake of certain parts of the environment. But nature too is often seen as bestowing gifts. The food we eat, our own biological selves, and the beauty we find in the natural world are seen in many cultures as gifts of nature, and in recognition of nature's fecundity, people wish to preserve the source of such gifts. These are familiar ideas, but neither of them are advocated by the theory. Rather the theory points to a third property—simply the "givenness" of nature (and not its capacity to give)—that will be taken up as a reason for seeing the natural as a moral category. The development of this argument though builds on other things that we will say about altruism.

While a geographical awareness points out that any one of these virtues can be the subject of gift-giving, it also shows that even as place may stress one, it interconnects it to others. The gift has consequences, and as moral agents we must take these into account. We must think ourselves into the places and contexts that our actions will affect, even if our actions are intended to be altruistic—even if they are inspired by intrinsic judgments.

Consider, for example, wanting to share a scientific truth or discovery. On the one hand it sounds good to say that truth should always circulate, but what happens if in this case the truth concerns a simple, readily accessible process for dispersing the deadly anthrax bacterium. The discovery describes how the use of chemicals found in any kitchen can be engineered to facilitate its spread. To circulate such knowledge will expand our awareness of reality and is a gift to the world only if it remains in the realm of meaning, but this "gift" is likely to jump realms, and become a powerful social tool with devastating effects. It "jumps" because this truth is related to the other strands in place. If it were given, it would be a gift that, while expanding our awareness of the real, diminishes the variety and complexity of that reality.

Clearly it is not altruistic to give something away that is likely to produce harm. So, anticipating the consequences of our actions is an important part of being altruistic. Thinking geographically helps us understand these consequences and applying intrinsic judgments helps us evaluate them. Certainly we intend to give without strings attached, but we must attach some such strings though they be only to increase the likelihood that the gift is good. We cannot simply give blindly, especially since a gift in one realm can spill over into others. And while this kind of assessment is counter to the idea of pure altruism, it is still the only way humans can approach it. We can be altruistic only imperfectly. It does not mean that the impulse is not there, or that it should not be cultivated. Rather it means only that it cannot be exercised completely and without contradiction.[19] Once again we see that intrinsic judgments both encourage this impulse and guide it.

Intrinsic judgments' guidance of altruism can be seen not only in alerting us to the interconnected of the virtues, but in distinguishing between what might be called a false altruism and a truer one. Immersing oneself in a cause, surrendering oneself selflessly and completely to its goals, may sound like an altruistic act, but it is false if the group itself is not guided by intrinsic judgments. This means that sacrificing oneself for the cause of the Mafia, of fascism, and even for certain forms of nationalism and religious fundamentalism are not morally worthy because these causes do not themselves adhere to intrinsic judgments. Intrinsic judgments require that we not only think of contributing to a group or project at one scale, but that we also understand how that entity to which we contribute in turn, contributes. Does it expand or contract awareness and variety and complexity?

Applying the moral theory guides our gift-giving, both in terms of inspiring giving and in terms of anticipating and evaluating its effects. Applying it also means making us aware that giving, receiving, and giving, even when the gifts may be principally from one of the realms, enter a web of geographic relations that combine and recombine them with elements of other realms.

Even so, it is important to examine each of the virtues separately in order to expand upon the pervasive effects of altruism on the theory, and to show how altruism rescues the virtues from relativism. I want in effect to restate what these virtues mean in light of altruism. In doing so, I want to set the stage for their use in the context of social relations that will be addressed in Chapter 7.

Altruism and the Virtues

Truth   Altruism is an essential part of this theory's view of truth. As we already noted, intrinsic geographic judgments see truth in terms of a relationship of correspondence to reality. The role of correspondence comes directly from the fact that the theory assumes the real exists and the good is real. While a correspondence theory of truth does not on its own presuppose that truth must be given openly and freely, doing so certainly increases the likelihood that we will come to better approximate the truth.

But in stressing a correspondence theory, intrinsic judgments also require that there be numerous places and points of view from which to see reality, and that all of these views be made available as public contributions or gifts—that they be open and accessible. The variety of views is essential because to know if we are seeing clearly we must be able to compare and contrast and sift and winnow from among them. Openness and accessibility—or an altruistic conception of knowledge—is essential because to keep knowledge secret is to diminish both the chances of knowing that it is real knowledge and not simply fantasy, and because to keep knowledge secret diminishes variety and complexity of the world.

So truth is not only expected to have some verisimilitude, but to be open and public—to be a gift or contribution, just like reality itself.

Justice   The theory makes an intimate connection between altruism and justice. The theory does not lead to detailed principles of justice that are very different from those found in most liberal democratic systems. Rather it provides different grounds for justifying them, and draws attention to how they are interrelated in a functional sense in the process of altruism. The core idea of justice is that people be given the opportunity to be involved in the creation of places and projects that allow them to both contribute to and receive the gifts of seeing through and variety and complexity. I say "both," but I intend this to mean that the purpose of receiving is to create opportunities to give. A central part of justice then is providing opportunities for people to make contributions and impart gifts. This, as we shall see (in Chapter 7), gives purpose to human rights, and inserts altruism into the basis of democracy and economy.

The theory does not stress one particular facet of justice, such as equality, need or merit, over others, but sees all of these as appropriate in cultivating awareness and in inspiring altruism. How the aspects of justice are mixed or emphasized at particular places or times is then guided by this imperative.

As with truth, which must be open and public, so too is justice. We cannot hoard it: rather it must be given, and given freely. Justice does not work when only a few have access to it. This is the basic idea of justice that stems from intrinsic judgments.

Rights How then do these principles of justice become expressed as rights? By way of a reminder, rights are political guarantees of particular types of justice. Without a conception of the good to support them, rights, like principles of justice, appear arbitrary. What is more, rights tend to be stated negatively, discouraging us from doing away with something, but not encouraging us to move in any direction. The theory's conception of justice leads to rights that are similar to those found in most democracies. That is, it supports the rights found in the Bill of Rights and the United Nations Charter on Human Rights (though as we shall see in Chapter 8, it does not necessarily support the right of private property). What it does do, as I noted above, is to show how these rights are necessary to, and justified by, participation in altruistic processes. It encourages us to see our rights as leading to moral obligation and purpose. This makes rights positive, rather than negative (a point I will explain in Chapter 7).

A more immediate issue about justice and the theory concerns the problem of elitism. How egalitarian can its concept of justice be if it is based on the capacity to contribute and to be aware?

Egalitarianism and Elitism The stress on awareness may suggest that the theory urges greater care and concern for those who have talents that enable them to see more clearly, or to be good, or to create gifts. Is this the case? Does the theory promote an intellectual and moral elitism? Would it not encourage us to value those humans who are more insightful and morally better than others? My answer is that at the most basic level it does not. But before I argue why, and how altruism plays a central role, I should point out that those who may be smarter are not necessarily those who are morally superior, and vice versa. But back to the reason why it does not promote elitism.

The theory imparts equal dignity to all reasoning creatures. Humans have rights and are treated as ends because of this. Even though people possess this capacity to varying degrees, they are all entitled to respect and care because they are such creatures. Even so, there is a boundary or limit of ability (though contested and shifting) below which a human loses some or all of his or her status. A brain-dead person, or a child born with virtually no brain and who will forever be unconscious, immobile, and on a respirator, is at or beyond the boundary of what we may call a human being. Even then, we treat that body with respect, though it does not house a responsible autonomous self, and most courts of law would honor requests from those responsible for such a life to expire. Drawing a line based on some lower limit in the capacity to reason may seem exclusive. But it can also be extraordinarily inclusive, for above the line can be any being that reasons, and not simply human ones. So, above this level, dignity and rights will and should be accorded regardless of further differences in ability simply because these beings can be responsible creatures. Their capacity to reason makes them moral beings, and one does not have to be very good at reasoning to be a very good person.

Still, it might be argued, that because the theory encourages us to think altruistically, those who have more talents in this direction will be those who are highly regarded. To this I can say that most moral and or social systems will likely value some individuals more than others. Even in Rawls's egalitarianism, those who are more productive and can raise the level of others are given higher rewards, and so would most likely have higher prestige. In our theory, however, the checks and balances of intrinsic judgments encourage a variety of things that can be contributions and does not measure them only in terms of the social and issues of justice, but as parts of the realms of truth and nature. Emphasizing place urges us to create a variety and complexity of gifts that are not possible to measure and compare.

An equally important quality of the theory that mitigates against elitism is that a focus on place draws attention to the fact that as we undertake projects we depend on the complex efforts of countless others. We are not simply individuals, but part of collectivities, working in and through place and dependent on the activities occurring in other places. Also, we participate in many places and projects and are the beneficiaries of many more. A geographical consciousness makes us aware of these interdependencies and thus leads to a sense that without cooperation, trust, and mutual respect, places and projects fall apart. In a hospital, for example, there may be a star surgeon who adds to its luster. It may seem that this surgeon is what the entire hospital stands for and, in one sense, that is true. Patients come there because of her, and she sets the standard. But, the surgeon requires a dense network of support, and so too do the patients and other employees, and excellence and care in doing this are part of gift-giving. An especially good nurse or an unusually competent and kind janitor can inspire others involved in this complex project. And when the surgeon leaves for the evening to find her sink at home no longer works, her kitchen needs the expert care of a plumber; and the surgeon's child needs the care given by a teacher; and both mother and child, the care given by friends, and so on.

In each case we may be able to single out the individual primarily responsible for an instance of giving and caring, but their efforts are dependent on a web of other individuals and places. And while the surgeon may have saved someone's life, a teacher may have molded the life of this child, and a plumber may have made her home life possible, and all may have done so with a degree of excellence and dedication that could inspire further examples of attention, care, and giving.

Inspiration and admiration are strongly attached to altruism. An effort to contribute, even if it is small and unsuccessful, still inspires and makes us attentive and grateful to any effort at all. It increases our awareness of the good, and that it can be present in the most diverse and smallest of details. Indeed, this is particularly evident when we consider aesthetic contributions. There, attention to the smallest detail can provide the greatest reward and inspiration to others.

Any gift, of any degree, if it is truly given, creates in the recipient a feeling of gratitude and of inspiration to do the same or even better—to give a gift to others. On the other hand, if a system stresses self-interest and instrumental judgments, then difference in talent and quality may produce emotions more along the lines of resentment and even envy. To avoid this, equality must be framed in terms not only of receiving, but of giving, and this is what the theory's concept of justice does as it considers it to be about equal opportunity for individuals to be involved in places that allow them to make and contribute, as well as to receive.

The natural Truth, justice, and the natural have always been complex and conceptually loaded concepts. The natural though presents problems of its own as a moral quality. Perhaps the most important comes to light when we think of the opposites or negativities of these virtues.

While the natural is often claimed to be a virtue—as when some argue that going back to nature will purify us and that nature should be preserved—we also find that resisting, destroying, and transforming nature is also often morally worthy, for, as the geographic problematic puts it, we can not accept reality, natural or otherwise, as it is. So it is a good thing to resist rays of the sun by building a shelter, or destroy the vectors of malaria through killing mosquitoes or controlling wetlands. On the other hand, we do not want to resist or conquer truth and justice; we really do not want to go against them, as we might against nature. How do we explain the relationship of these virtues and the particularly ambivalent qualities of the natural?

Two points help clarify why the natural is valued, but ambivalently. The first stems from particular historical twists in the problematic. If we return to the geographical state of nature thought experiment, we find that we are first geographically naked in and surrounded by reality or space, which can be called nature. This nature is untamed and mostly threatening. We create places in order to escape it and transform it into what we think it ought to be. The places we create provide us with some order, predictability, and control. They also increase our sense of being apart from nature.

As we come to realize that this wilderness seems to be disappearing and that we no longer live in a few small-scaled places surrounded by a sea of space or nature, but rather have now made virtually all of earth space into place, and if we are not happy with it, we may decide to escape back into space or nature. We now find nature appealing in part because we are not only estranged from it, but because we have, or are threatening to, dominate it; and, as Tuan points out, dominance can lead to affection: we become fond of things we control.[20] The original state of nature was not something about which we were overly fond because it threatened to dominate us. Now we think we can turn the tables and so come to treat it with the affection of a pet. But, this return to nature is an illusion. No matter how much we yearn to, we cannot live without place, and the "unbounded" nature to which we want to escape has now itself been made into place; wilderness areas and even Antarctica remain mostly nature because we have demarcated them as places whose rules are intended to restrict our own interventions. We may not want to admit this, and make it seem that when we enter such places they are really unbounded spaces (and represent them as such in literature and art); yet the fact is they are now bounded places with humanly constructed rules that affect what takes place. And even then, we can be in them only temporarily. To stay longer is to transform them even more. So we must again escape nature, even when it is a place, and return to our other, more cultural, ones.

These historical turns in the problematic can explain why it is that nature-as-place seems to have a positive and even purifying quality in the contemporary world. Indeed, some may frame this escape from culture as an escape from the profane into the sacred, as when individuals enter the wilderness expecting to become purified and to commune with something spiritual. But this explanation is far too contingent to support the arguments of the theory. That is, there is nothing in this narrative to prevent us from changing our minds about nature. We may come once again to value morally the places we create and tolerate or even encourage an increased interthreading of culture with nature. This narrative is also too narrow because it considers only external, and not internal, nature. That is, it does not explain why we may sometimes want to tame our biological selves and other times revel in it.

The second point about nature as a virtue is more basic and general, and returns us to the altruism and gift-giving theme of the theory and is the one the theory employs. The argument is that no matter how places intertwine the threads of nature with those from other realms, and no matter how much of reality is filtered through place, we are still able to recognize that some threads eventually lead back to a realm that is not humanly created or controlled and to which we must bow. These components amount to the idea that nature is there as a *given*. These qualities of *givenness* share similarities to the idea of gift-giving. Nature, in its givenness serves as an inverted image of gift giving. On the one hand, it provides a glimpse of something like selfless giving, but on the other draws attention to the limits of pure giving, and its impossibility for humans.

Gift-giving for humans requires a purposeful gift-giver—a person who gives. This clearly is not what I think nature does in its givenness. I want to state emphatically that I believe that nature has no purpose, plan, or intent. Moreover, its acts or effects can be harmful as well as beneficial. So, again, without a purposeful giver, and with "gifts" that often have negative effects, how does nature enter the gift-giving picture? The answer is that its qualities of *givenness* have the capacity to reflect our idea of complete altruism into an image of what the world looks like when it gives without purpose and intention; but in this image we also see the impossibility, and even undesirability, of humans achieving such a state, for giving without intention and limit for us can lead to a givenness that is not good and that can destroy both the giver and receiver.

Consider some of the comparisons and contrasts. In addition to the givenness of nature that reminds us of a gift, there is also the fact that gifts given by humans are often described in terms that remind us of nature, as when a genius and a caring and kind person may "bubble over" with the gifts of their ideas and kindness. These gifts are not intended to have strings attached; the recipients can accept or reject them, and use them in turn to create something new. Recipients often do not know the origin of such gifts. The inventors of the wheel or the decimal point are anonymous, but their inventions are there for us to use—as givens. And so too is our everyday language. And there can never be enough givens or gifts. We can never have enough knowledge and variety and complexity (in their jointness.) So, like nature in its givenness, our gift-giving strives for plenitude and continuous circulation. But unlike the givenness of nature, we humans must not release simply anything into the world and call it a gift. We have to think about its consequences. So for us gifts are purposeful even if we do not control how they are used. And each of us personally cannot give unboundedly and continuously. We are limited and must give strategically. If we do take into account our own self-interest to some degree, we destroy ourselves and are no longer able to give at all. So in its apparent plenitudinous givenness, the mirror that nature provides reveals in its vanishing point behind the surface the act of a purely disinterested and selfless givenness—the asymptote of pure givenness—but this is not an image of ourselves.

This of course is an interpretation of nature, but it is not an anthropomorphization. It becomes so only when we push aside this idea of an image of nature and come to believe that it really has purpose and really does provide gifts in the sense that humans do. Nature can and has been valued for a host of other reasons. What I am saying here is that this particular role of nature as mirroring and thus inverting pure altruism is the only way the value of the natural can play a role in altruism in particular and in the theory in general.

Locating nature's virtue in the gift explains why the natural is an ambiguous virtue and how it is related to the others. It also points out what our relationship to nature ought to be. We have an obligation to ourselves (and to future generations) not to diminish those elements of nature that can be used to create an expanding reality and awareness of that reality. This applies to both external and internal nature. Seeing nature in the context of place, and under the guidance of intrinsic judgments, defuses the question of knowing exactly which elements are purely or really natural and which ones are not, and also defuses the issue of trying to set these apart from the rest of the world, as in wilderness areas or nature preserves. Nature, along with social relations and meaning, must be woven into place which then hybridizes them. And if intrinsic judgments are correct, following them can help assure that we do not jeopardize our chances of creating places that enrich the variety and complexity of the world and our capacity to understand it.

The loom of place allows us to weave infinite varieties of truth, justice, and the natural; and intrinsic judgments encourage this variety, while not allowing the virtues to become relative. This is due to the regulative qualities of seeing through to the real and variety and complexity and to their connection to altruism and gift-giving. While intrinsic judgments encourage altruism, they also guide and check it. They enable us to understand what a good gift can be, and it shows that not all forms of altruism are good. It indicates how we can know if we are giving ourselves freely and altruistically to the wrong or the right kinds of projects. Even then, we can encounter difficulties. One of these stems from the fact that the acceptance of intrinsic judgments does not guarantee good effects. There will always be unforeseen events and unintended consequences. Also there is no formula for balancing variety and complexity with seeing through to the real. Indeed, there is no formula for measuring either. It is up to our best judgment to decide, and judgment is always controversial and fallible. Still, it provides some degree of guidance that is grounded in a larger picture of what ought to be.

Assuming that intrinsic judgments can help, there is another problem that arises from the quality of our place-based nature. At the same time that we want to be made more aware, places also disconnect us. They shield us from much of reality so that we can focus our attention. This can be for the good if the result is that we can then make a contribution, but it also can be for ill if it is used to isolate, deceive, and distort. The capacity of place to connect and disconnect, to help us engage and withdraw, has a great deal to do with our chances of being good or evil. We must consider these issues in detail, especially how the compartmentalizing qualities of place can readily lead us to self-deception, which is the locus of moral drift and evil.

6

GEOPSYCHOLOGICAL DYNAMICS

THE THEORY OUTLINES HOW THE STRUCTURE and dynamics of place can help or hinder our attempts to be good. Here I want to focus especially on the role of boundedness and boundaries, for they encapsulate the effects of many of these dynamics, and because they join neatly with the psychological process of compartmentalization. The boundary of place can connect and disconnect, making us part of, and apart from, the world. I will explore these relationships and their moral implications mostly at the personal or psychological level. I will refer to this geographical boundedness as compartmentalization because the term is also used to describe building boundaries in our minds.

The point is that psychological and geographical compartmentalization are interrelated. At the personal level we can oscillate from moment to moment between using geographical compartmentalization to connect us and expand our vision, and other times using it to disconnect, and not attend to what we should. These personal oscillations are in turn affected by the actions of others and the society as a whole. If we are in a good society, it is easier to continue using geographical compartmentalization to expand horizons and concerns than to narrow them, but if we are in a bad one it is easy to use it to narrow them, and requires heroic effort to do the opposite.

These geopsychological dynamics are complex and will be addressed in three sections. The first is an abstract overview, which begins with morally neutral uses of geographical compartmentalization, and then considers compartmentalization that moves us along the continuum to a more situated, narrower and eventually morally irresponsible position that take the forms of moral drift, self-deception, and evil, and then discusses the difficulties and possibilities of reversing direction.

The second section revisits two of the three cases discussed in Chapter 4—that of the antebellum South and Nazi Germany—to illustrate the geopsychological nuances in the use of compartmentalization to create moral drift and self-deception. First, I take the case of the antebellum South, where I use Thomas Jefferson's struggles with slavery to suggest how difficult it is even for a person with his intellectual gifts and morally good intentions to see clearly and do the right thing, and to convince others to do so, when moral drift and self-deception are the order of the day. The Jefferson case stands as an example of the effort required to begin to turn the negative effects of compartmentalization around. The case also stands as an example of the power of the realm of meaning and truth: Jefferson's influence in the issue of slavery is due less to his actions and more to his writings and philosophy. Still, Jefferson could have tried harder and done more. In the second case—Nazi Germany—there were many who did resist both through actions and ideas in ways that were more heroic, for they risked their lives (and the lives of their families) and not simply their reputations and fortunes.

I leave these heroic actions for the third section, because I want to have the rich documentation of the Nazi example fill in the details of how geographical compartmentalization is part of massive self-deception, not only for the populace overall, but for Nazi leadership, and for Hitler in particular. Hitler is the key. I want to show that if we hold him morally responsible for his actions—that is, if we do not think he was mentally insane, and could not do otherwise, and so is not responsible for what he did (that is, if we do not accept an insanity plea from some hypothetical defense attorney of his)—then self-deception becomes the central element in explaining his actions. Hitler extoled the virtues of a closed and unquestioning mind. This is in stark contrast to Jefferson, who not only valued enlightenment, but struggled to become enlightened. True, the actions of both men perpetrated evil (Jefferson was a slaveholder throughout his life). But it is fatuous to suggest they are the same. The harm Jefferson did was confined to his slaves, and in significant ways countered by his public pronouncements. He condemned slavery, and his efforts to become more aware and to struggle against moral drift and self-deception are precisely the qualities that must be emulated if we are to do better.

Qualities such as these are taken up in the third section, which focuses on the use of compartmentalization for the good. I discuss here the actions of good people resisting evil, and also the capacity of doing things well and aesthetically as a means of moving us in the right direction. Finally, we must be reminded that the compartmentalizing qualities of the boundary of place are the primary theme of this chapter, in part because they fit the psychological idea of compartmentalization. And while it is easiest to think of compartmentalization as following from the rules of in and out, it in fact involves all of the structure and dynamics of place.

GEOGRAPHICAL AND PSYCHOLOGICAL COMPARTMENTALIZATION
Compartmentalization and Self

Drawing on a wealth of ethnographic and archeological information, Peter Wilson, in the *Domestication of the Human Species*, argues that a major watershed in human history was the development of relatively stable and permanent partitions or walled-in structures within a community or settlement.[1] They were revolutionary because of how they allowed people to be part of a community and yet apart from it—connected and yet disconnected. Before their introduction, a human band or community would live either in a cave or in the open (or use partial and temporary enclosure such as a lean-to), and every member could be seen by everyone else. Even when the band or community contained several households or subgroups, each seated around their own hearth, the collective members would still be assembled together, so that one could turn around and see the members of the other households. Privacy was possible of course. Eyes could be averted, and people could try not to listen. But to be truly alone, one had to step outside the confines of the cave, or go beyond the range of the hearths, and into the bush. Remaining there too long could raise concern or suspicion. For when out of view, the individual could do things that were out of the ordinary and even threatening to the rest—perhaps the person could be encountering strange spirits, or casting spells; the community was reassured when that person returned and resumed his or her place and so was again open to view. When cultures developed more permanent partitions—or when humans became "domesticated" as Wilson puts it—the lack of visibility that these segmented places provided brought the possibility of the unknown and the suspicious behavior that accompanies it directly and permanently within the community itself. In other words, with such compartmentalization and opacity comes the possibility of undertaking new projects that may not be visible, knowable, and consonant with those of others in the community.

With the dawn of domestication then comes a tension between connection and disconnection, even within a group. Spatial compartmentalization also leads to complexities in defining one's self. The community as a whole, when open to view, is more a single unity. True, there are parts, such as households gathered around their own hearths and individuals who compose them, but the degree of distinctiveness is lessened by the lack of geographical segmentation. With compartmentalization, we find not only an increase in the differentiation of groups into subgroups, but an increase in the sense of self.[2] This is the strand that Tuan explores in his *Segmented Worlds and Self*.[3] Take a hypothetical society that is already somewhat domesticated in Wilson's sense, where its members and I are almost always together. Certainly, the pronoun "I" makes me somewhat distinct or separate from everyone else, but my identity is still driven mostly by affiliation with the group as a whole. I am constantly in their presence, seeing, hearing, and touching them. But now imagine that the society has become slightly more complex so that I now must engage in numerous projects in different places, with only some members of the group present. Even though I am not alone in any of them, I still may become more aware of myself as an individual with a distinct biography, because being in one of these places allows me to express a certain facet of myself and also to distance myself from the projects and places of others. So this additional spatial compartmentalization or segmentation could lead to an increased sense of self-consciousness, or sense of self. This can also happen when I am alone in a place. Compartmentalization may now be complex enough so that I have my own room in my family's hut, or even my own hut. These would provide places to which I could withdraw out of view from everyone else, and so again I could become to some degree detached and differentiate myself from the rest. This process of compartmentalization could progress even when the boundaries are not literally walls keeping us from viewing others. I might have a particular place in a room, or a seat on a bench, or in a more segmented way, a separate chair. Tuan shows that as societies become more compartmentalized in all of these senses—or as places multiply and are thinned out and specialized—the members of the societies develop a fuller and deeper sense of self. All of these processes can be summarized in the case of eating.

In a simple society we may kill game and bring it back to the camp where we sit on the ground side by side around the fire ripping and carving pieces of meat from the joint roasting on the spit. In a more segmented, but still simple society, a peasant culture perhaps, the source of meat may be domesticated. It can be part of a shepherd's flock, or raised in a farmer's barn. The animal may be transported to a butcher, who slaughters it, and then brought home, where it is cooked over a fire in a kitchen and then consumed in a common room. Different places/projects/specialists segment and compartmentalize the process. Those who raise the animal may know only vaguely how and where it will be consumed, and those consuming it may similarly have only a vague idea of the stages involved. Each of these stages or processes, even in a relatively simple peasant culture, can be amplified—the geography of animal husbandry, the different territorial organizations attending shepherding and farming, all can be explored. The process of transportation can be examined; attention can focus on the butchering processes, or on the details of preparation and consumption of the animal in the home. In each case ever smaller territorial units/roles/projects can emerge. A self that is involved in one of them acquires attributes that differentiate it from another, and enables one to distance onself from others to become more self-aware. In a complex modern society the territorial units, roles, and projects involved in the process have multiplied enormously.

As these examples multiply, so too does the compartmentalization that occurs even within one place. In the simplest societies, we may all be seated around a fire and eat with our hands. In a peasant society we may be seated with other members of the family on a bench which is alongside a table. The food may be placed in a single bowl in the middle of the table, and members of the family take portions of this food onto their own plates. Do they have spoons, knives, and forks? Do they have their own cups and glasses? And what happens if instead of benches we have chairs, where each person sits apart from another, and what happens if different types of food are placed on different dishes? From sitting around a fire to sitting on benches, to sitting on individual chairs with our own place settings and separate serving dishes, moves us further into a spatially segmented life, which, as it occurs even within a single room, has me farther removed from the food, in that the utensils and plates distance me, and from others, for I have my own chair that sets me apart. I become more individuated, in the sense that I am more self-conscious of how I eat, of where I am, and of myself in general.

The same trends can be found in every other aspect of life. Drama and music, for instance, in simpler societies were collective events occurring perhaps around the camp fire. The entire group may have participated in songs, dance, and the enactment of stories. In contemporary times, these occur for the most part in separate places—in theaters—that may in fact specialize in very distinct types of performance. Within these theaters one usually finds performers and audience separated: the performers on stage, and the audience seated, each member in his or her own chair. With the lights dimmed, a member of the audience may not even be aware of other members, but only of himself or herself and the activities taking place on stage. We can toy with these relationships and try to recombine what we have segmented. We can have audience participation, "take theater out into the streets," have everything become performance, and so on, all the while having these new configurations become commentaries on the other, more segmented form, just as we can decide to have benches replace chairs in a dining room, share a pizza that we eat with our hands, or better yet, have a picnic or eat and sing and dance around a campfire. But these are now options that arise because we have already segmented the world enough to create selves that are faced with choices about their own identities. Often we try to reduce the isolating effects of segmentation through modern forms of technology—cellphones, the Internet, television now pervade our own places so that we are never "alone"—but these allow a self already aware of its identity to feel reconnected. A self that wishes not only to become aware but to also reflect and weigh alternatives, and to do the right thing, often needs the opportunity to withdraw from the world. This can be done simply by willing it—by forcing oneself to push all else aside and concentrate—but it can happen with greater ease by having a place to which one can withdraw. Critical distance and self-awareness are linked overall to spatial compartmentalization. This is one of the ways place helps constitute the raw psychological materials for doing good or evil—in having a hand in molding the agent or the self.

Compartmentalization and Escapism

Spatial compartmentalization can promote a sense of self and a critical distance toward others and the world. But whether this is used to move us in the direction of instrumental judgments that tend to disconnect and narrow our vision, and reconnect us only when and as it is important to particular projects, or whether it moves us toward intrinsic judgments that encourage us to have an open mind, depends on where along the continuum the place is already located and on the exercise of our own wills.

The same can be said of the related effect of compartmentaliza-tion—escapism.[4] Places allow us to escape from reality into another reality. In terms of the problematic this means that, instead of saying we cannot accept reality as it is, we can say that we need to escape from it.[5] We can do so through creating and changing place. Escapism ap-plies to all of our previous examples of compartmentalization. When we build a hut or lean-to, we are providing a way of escaping from the rain or the sun. When we erect partitions and walls, we are escaping from the view of others. When we stretch out a process through vari-ous stages and places, we can be escaping from an awareness of the whole. The moral issue is whether this escape prevents us from seeing reality and its interconnections clearly.

Escapism can be good or bad. I hardly think that escaping from the weather means we do not want to know about it; nor do I think stretching out a process into many segmented steps means that we are not interested in how the whole is interconnected. There are certain things most human beings would rather not confront, and often these are related to our biological selves. We often want to avoid facing the fact that we are part of nature—that we are partly animals. Using com-partmentalization to escape our own animality occurs in several ways. Urinating, defecating, eating, and having sex are all things that we share with other animals. But they are also things about which we are partic-ularly self-conscious. We usually do not want to be observed urinating and defecating. We relieve ourselves away from others, and often in places designed especially for these purposes. Most of us do not want to be observed having sex. This also almost always takes place away from others, or in the privacy of some enclosed area. Even a curtain around a bed in a room shared by others can provide some degree of privacy. Of course we can find exceptions to these generalizations, and many can "make a statement" by wanting to be observed. Some may have particu-lar sexual fetishes and enjoy being seen sexually or enjoy seeing others, but overall, we seek privacy when we do these things.

Eating too is part of the process of escapism, but in a more complex way. Eating is usually a social act. We want and even crave company. Eating is a means of binding people together. But we do not want to know about everything that supports a meal. We usually try not to rec-ognize that eating is a process that involves destroying other things so that we can be nourished. Vegetables are alive. We pick them and then eat them. But "picking" them is a term that disguises that fact that we are dismembering and even killing them. We want to avoid being reminded that our lives are sustained by the sacrifice of other life, even vegetative life. This is more so with animal life. As we are eating meat, we may not want to be reminded of the fact that animals may have suffered for our needs. Cooking itself is a form of disguise. Sliced meat with gravy may seem only remotely connected to a particular part of an animal. Even our words for meat can serve as a means of disguising its origins (and also elevating its status). When we eat meat from a cow or a steer we call it beef. The meat of a calf is called veal, a deer is called venison, and so on. And the series of spatial partitions—from farm to slaughterhouse, to supermarket, to kitchen, and then to dining room—all add to this avoidance of making connections. Still this form of escapism is not necessarily bad. It does not mean that we do not eventually want to put things back together, and confront our own animality. Rather we simply do not want to be reminded of it on all occasions and all of the time. This is most especially true of our own mortality. Because of our ability to symbolize and think abstractly, we may be the species that is most acutely aware of death. (Indeed other animals, though they avoid it, may not be aware of it at all in the sense that we are.)[6] Yet we try very much to escape our own mortality. We do so by hoping to "live" through our progeny, through our contributions, and even to live on in some sense of immortality—through being reincarnated, or entering an eternal life in heaven. This is not necessarily irresponsible, as long as we recognize that an end in some form cannot be escaped, and that it lends weight and meaning to life.

Indeed, degrees of compartmentalization and escapism are necessary and healthy. If I were overwhelmed by the fear of death, I may become immobilized. This would be true too if I were to be exposed to everything about the world. I need to escape, at least to some degree, in order to focus my attention and to undertake projects. I need to escape simply to relax. Even escape to a fantasy world can be helpful. I could do this by closing my door and watching a soap opera on my television; or I could do this by entering a landscape designed for escape and fantasy. A game arcade or a Disney World is such a place. Indeed, many places of consumption—shopping malls, tourist sites, and vacation spots—are there primarily for escape. Are these bad places because they do not encourage us to see reality clearly? That depends on how deceptive and pervasive they become. Intrinsic judgments help us decide how to weigh these and all of the other tendencies of compartmentalizing the world—both as compartmentalization obscure reality and yet focus our attention upon it—and help us to know if the balance is tipped in the right direction. Here though I would like to consider what happens when these escapist qualities of compartmentalization move us in the wrong direction.

MORAL DRIFT AND SELF-DECEPTION

Compartmentalization can have the effect of narrowing our vision. When this is not countered by an equal attempt to expand it, we find ourselves moving almost imperceptibly and inadvertently along a path which Laurence Mordekhai Thomas calls *moral drift*.[7] Moral drift, because it does not require effort on our part, is always a drift downward or toward the instrumental end. It can begin most innocuously, for, as we noted, place's role in compartmentalizing helps focus our attention on projects by cutting us off from the rest of the world. The point is that we can begin this drift if we do not then make an effort to reconnect, not simply as dictated by our projects, but by our commitment to make a contribution to the understanding and creating of a more varied and complex world.

When we become less aware and more inward looking, we may subtly narrow our sense of moral obligation. We may no longer make an effort to overcome a complex and isolating world. The segmentation of place can be used to keep us from caring about anything else that does not fit our own projects and interests. In turn, we create places that then promote this more narrowed view. If I accept the limits placed on me by the geography that supports particular instrumental projects of mine, if I take these as my universe of concern, and if I then support only the places that encourage this, my moral principles and actions will drift downward and away from the good. I still may not hate or do evil. I may even admire strong and universal moral principles. But if I do not make the effort, then in spite of what I may claim are guiding moral concepts, my own actions and principles will have diminished moral scope. I will be motivated primarily by self-interest. What happens here for me happens also for others. The collective effort to see through to the real and to create variety and complexity diminishes. As this happens there also will be fewer institutions with an altruistic bent that will receive support, and so we drift down a moral spiral, making it ever more difficult for an individual to be good.

Abandoning conscious efforts to expand one's obligations, and settling instead for the less ambitious position of simply holding that a person should do no harm, is the beginning of moral drift. The first step is a move from being good to perhaps being decent.[8] So, in a segmented world, I may use compartmentalization to not have to encounter the poor, or fight against censorship or environmental degradation; and why should I, if I have not directly and intentionally created these problems? I never intended to impoverish anyone. Nor have I ever promoted lying or have intentionally polluted. Rather I have obeyed the laws and rules of conventions in my daily life. So I am not responsible for these conditions. If someone else wants to do something about it, then well and good, but I do not have such a moral obligation. I may still admire those who want to do more—who may have higher moral standards (if I can in fact still recognize them)—but I and those around me expect less of ourselves. This substitution of decency instead of morality, this movement toward moral drift Lawrence Mordekhai Thomas calls common sense or laissez-faire morality. Under such a system I may perhaps offer a helping hand, but only if it does not overly inconvenience me. In such a world, with such an attitude, I may still shudder when I hear racist remarks or wince as I see someone pouring toxic chemicals down the drain or pity the poor and the homeless, but since I believe I did not create these problems, I also believe I am not responsible.[9] I am now using compartmentalization to allow my moral vision to drift to a narrower and more instrumental position.

I do not mean to say that this decline would necessarily occur in all directions at the same rate. Some strands of truth, justice, and the natural may be more strongly supported than others. But without a conception of the good, these strands become interthreaded in places and projects that are now guided almost entirely by instrumental goals, and we have seen how this then distorts strands and stamps them with their own meaning. Something like this has happened to many ordinary individuals in bad societies. I do not mean that this explains it all—for it ignores, though it can be connected with, ideologies of hate and racism, which require that a large number of people do nothing to resist them—but it does helps us understand how a people can come to live morally diminished lives that they still think of as good, or at least decent, and how in so doing they could simply through indifference and neglect allow others to suffer and not prevent some from inflicting immeasurable pain.

Geographic compartmentalization not only provides an excuse for moral drift, but can provide an environment that makes one even unaware that drift is occurring. Having some things in and out of sight and those in sight conforming to particular rules of place creates a sense of how things are, or are supposed to be, by "nature." A large slave plantation is a complex geographical system that segments the races, and reinforces stereotypes. The whites in the main house were served by black slaves, who lived in substandard quarters. The rules of place and spatial segmentation made slaves behave slavelike, which to the whites confirmed that they were not fully human. The same is true with the ghetto and the concentration camp. The rules of place make stereotypes appear real, persistent, and natural, and compartmentalization allows one to not have to face the contradictions they entail. Compartmentalization can be used to select our "moral audience," or it allows us simply to be placed within one, and not question it. "Consider the difference between wrongdoing in complete privacy and wrongdoing that others witness. The latter is clearly more acute, if only because one's moral weakness is bared before others; in the former case, one can take comfort in the fact that no one knows. Ironically, however, wrongdoing in the company of others may sometimes provide even greater comfort, for then it can be said the others were not able to resist either."[10] Being in places that are accepted uncritically promotes a common sense or laissez-faire morality—a moral drift— that does not inspire us to draw this audience and current practices into question. Rather it takes the instrumental uses of places as the definition of the moral. This allows many to believe that their individual actions were not creating evil, or doing anyone harm.

Moral drift and self-deception are interdependent. Moral drift is about social and psychological conditions of compartmentalization. Self-deception is about the psychological. Its focus on the self is why it plays such an important role in the theory. How then does moral drift lead to self-deception? In order to do justice to the shift, I need to recapitulate the process so far, starting with the fact that compartmentalization is a part of life and can lead to good or evil. On its positive side, it allows us to focus attention and to undertake worthy projects, even as it is narrowing or concentrating our vision. Consider that I am interested in eradicating poverty, but may not want to confront the poor in my city. I may avert my eyes from the panhandlers on the street, and not stop to help a homeless woman find her way to a shelter. A positive interpretation of these actions is that I may avoid this confrontation of poverty and engagement with it because I want to save my limited abilities to focus on what I can do best for the well-being of others. This avoidance then is not moral drift, or self-deception. It may be an accurate appraisal of my capacities. That is, if I attended to every problem, I would be exhausted. I must narrow my vision to undertake even a moral project. I am not deceiving myself because facing what I have tried not to see will not threaten to undermine the validity of my assumptions. Focusing attention then is not necessarily self-deception. Not wanting to be reminded of my own inabilities to cope does not make me less aware of them, just as not wanting to be reminded of my own bodily functions, and hiding them from the view of others does not mean that I am not aware of my own animality. Reminding me of these facts does not change my overall understanding of my own capacities and limitations.

When compartmentalization diminishes one's moral commitments, we then enter moral drift. But this too is not necessarily an act of self-deception. I may not want to know about the poor and the suffering in the world. Instead of thinking about it at all, I will turn on the television and lose myself in a soap opera, or go to an amusement park. I want to escape, pure and simple. I am not denying that there are others who need help. Nor am I deceiving myself into believing that I am doing the best thing—that I am a truly good person. But I may believe I am not doing the wrong thing either. I am not doing evil. Rather I think of myself as a decent or at least an ordinary person who does no harm to others himself; and that for me to be a good person would require an effort that I cannot make, or if I can it will be almost a superhuman effort on my part and so I will wait until it is absolutely essential.[11] And if others feel the same way, then this use of compartmentalization makes me no worse than the others around me. As others come to think this way, then for the culture as a whole, this use of compartmentalization not only lowers our moral expectations, but also limits our awareness. Facts we should know are no longer taken into account.

Unlike focusing attention and moral drift, the issue of self-deception arises when we use compartmentalization to actually deny the existence of a set of facts. It occurs when a person holds two contradictory types of evidence in mind—one set supports "p" and the other its opposite, "not-p." Even though the evidence for "not-p" is far stronger than the evidence for "p," I will avoid calling upon "not-p," or will keep my knowledge of "not-p" compartmentalized and apart from my knowledge of "p." I won't let myself confront the fact that "not-p" contradicts "p." Compartmentalizing "not-p" so that I do not need to confront it is a result of a weakness of will.[12] I am mentally and geographically compartmentalizing the world in order to not see clearly. In not mentally confronting the evidence, I am also not physically placing the evidence before myself and conversely, by keeping the evidence from view, I do not have to confront it and forsake my belief in "p." Even though self-deception is different, it can be abetted by, and built on, these other uses of compartmentalization.

How then do we get ourselves out of this? If the rest of society is open, and presents abundant cases of "not-p," it may not be difficult to eventually see it and abandon "p." But if the society is closed, then what? Even under conditions of moral drift and downright evil, it is possible to become less situated and to question what is taking place, and also to resist it by establishing different and better places. But this requires unusual moral character and effort in a system that conspires against it. And it requires that the rules of place and its compartmentalizing role be used differently. How this can be done will be the topic of the third section. But for now, since the role of self-deception is so important to the theory as an explanation of evil, we must consider more concrete instances of its dynamics and the difficulties of extricating ourselves from them. This is the purpose of the next section where we will again examine the cases of slavery and Nazi Germany—both standing as examples of massive self-deception. In this discussion I will make the important contrast between Jefferson's struggle to not be self-deceived and Hitler argument that self-deception is a virtue. The Jefferson case and slavery are discussed first (which reverses the order of Chapter 4) because slaveholders in the antebellum South were born into the system: they did not create it and in this sense it was easier for them to take it for granted and be deceived, with the exception of people like Jefferson who tried mightily not to be. In contrast, the Nazis and especially Hitler planned the landscape of self-deception and were so morally weak that they succumbed to it.

SLAVERY AND JEFFERSON—
NAZI GERMANY AND HITLER

Slavery and Jefferson

Elite pre-Revolutionary plantation owners, including Washington, Madison, and Jefferson, generally agreed that slavery was evil. This group offers an interesting case of those who are perpetrating evil, yet wanting, and trying, not to. They attempted to avoid moral drift and self-deception. The writings of Jefferson (and those to him and about him) on the subject of slavery show the amount of time he devoted to this effort. Though he did not in the end act as heroically as he might have and was subject to some degree of self-deception, his primary efforts, especially his intellectual contributions in challenging the narrowing and self-deceptive qualities of the landscape and the ideology of slavery, and of condemning slavery in general, are worthy of attention. They illustrate how complex and multilayered are the psychogeographical dynamics; how important it is to not only do the right thing, but to present an argument for it so that others can be inspired to do what is good. He was not as good as he could have been, but he was aware of the difficulties of self-deception and therefore made extraordinary attempts to avoid it.

As a slave owner Jefferson most definitely had an instrumental stake in the system. Yet his early pronouncements on the subject were unambiguously in opposition to slavery, and even though in later life he became a less active opponent, his state of mind was never at ease on the subject. His most uncharitable critics condemn Jefferson's position on slavery because he was not against it enough: "The test of Jefferson's position on slavery is not whether he was better than the worst of his generation, but whether he was the leader of the best . . . Jefferson fails the test."[13] While this may be true of the later Jefferson, it is not of the earlier man, who was near if not on the top of the list of critics against slavery. He denounced slavery as an evil in his "Notes on Virginia" and proposed in the 1780s a gradual abolition of the system, an end to the slave trade, prohibition of slavery in all the Western territories, and emancipation of all children born of slaves after 1800.[14] As Joseph Ellis puts it: "Throughout this phase of his life it would have been unfair to accuse him of hypocrisy for owning slaves or to berate him for failing to provide moral leadership on America's most sensitive political subject. It would in fact have been much fairer to applaud his efforts, most of them admittedly futile, to inaugurate antislavery reform and to wonder admiringly how this product of Virginia's planter class had managed to develop such liberal convictions."[15]

Still it is true that in his later years, Jefferson's resolve weakened. He no longer actively pursued emancipation, leaving it to the next generation, which he hoped might be more enlightened and thoroughly imbued with the spirit of the American Revolution and the ideals of equality. He also worried about what would happen after emancipation, for the pain and suffering caused by slavery could not easily be erased. This troubled him even in the Notes, where he saw a just God siding with the African Americans in retribution for this evil institution: "Indeed I tremble for my country. When I reflect that God is just: that this justice cannot sleep for ever: that considering numbers, nature and natural means only, a revolution of the wheel of fortune, an exchange of situation, is among possible events; that it may become probable by supernatural interference! The Almighty has no attribute which can take sides with us in such a contest."[16] Concerns such as these, as well as a growing political realism and a desire to push these issue aside for a future generation, led him to conjure up even more obstacles and unworkable solutions so that toward the end of his life Duc de La Rouchefoucauld-Liancourt was able to portray Jefferson's internal tensions this way: "The generous and enlightened Mr. Jefferson cannot but demonstrate a desire to see these negroes emancipated. But he sees so many difficulties in their emancipation, even postponed, he adds so many conditions to render it practicable, that it is thus reduced to the impossible."[17]

Detesting slavery intellectually, and yet not continuing to advocate its immediate abolition, is not an example of taking the highest moral ground. Compared to his earlier pronouncements, his later position is touched by a moral lassitude or drift. Such drifting was reinforced by the geographic conditions of his plantation. According to Joseph Ellis "The organization of slave labor at Jefferson's plantations reinforced this shielding mentality in several crucial ways . . . his cultivated lands were widely distributed, half of them at Bedford, several days' ride away . . . [and] Jefferson seldom visited those remote estates . . . he seldom ventured into his fields at Monticello or Shadwell . . . . When Jefferson did encounter [his slaves] . . . , it was usually . . . where he supervised them as workers doing skilled and semiskilled work [i.e., in a nail factory and as masons and carpenters rebuilding his architectural dream at Monticello.] All the slaves in the household, and most of those [nearby] . . . were members of two families that had been with Jefferson since the earliest days. . . . They enjoyed a privileged status within the slave hierarchy at Monticello . . . and were part of Jefferson's extended [and some most likely part of his actual] family."[18]

What is most striking is that they were also mostly light skinned. La Rouchefoucauld-Liancourt commented in 1796: "I have seen . . . , especially at Mr. Jefferson's, slaves, who, neither in point of color or features, showed the least trace of their original descent; but their mothers being slaves, they retain, of consequence, the same condition."[19] The geographic organization of Monticello disguised the full import of slavery. "In short," according to Ellis, "Jefferson had so designed his slave community that his most frequent interactions occurred with African Americans who were not treated like full-fledged slaves and who did not even look like full-blooded Africans because, in fact, they were not."[20]

These are examples of compartmentalization that encourage a degree of moral drift. But do they also involve self-deception? Did Jefferson geographically and mentally compartmentalize so that he did not have to face facts that would have required him to change his beliefs?

Ellis seems to believe this might be the case. Jefferson's behavior on slavery (and a host of other issues) "was . . . more [like] self-deception . . . [for it allowed him to] walk past the slave quarters on Mulberry at Monticello thinking about mankind's brilliant prospects without any sense of contradiction."[21] While I believe there is a point to this, it is not a very strong one. The issue here concerns how much of reality is then ignored and how unreasonable are the beliefs that are being shielded from that reality. Certainly Jefferson may have used this geographical and mental compartmentalization to suppress some facts and to create some illusions, such as the belief that he treated his slaves with respect and as though they were members of his family, when in fact they were slaves; or the idea that because of (what he believed to be) his enlightened treatment of his slaves, he should then be forgiven if he no longer makes as great an effort at emancipation as he once may have; or, as we shall see, his granting of the moral equality of the races, but then dwelling on the dubious distinction between the intellectual differences between them. These may well be instances of self-deception, but there is no evidence that they ever overshadowed Jefferson's vision of the bigger picture—his conviction that slavery was an evil that should be abolished.

Jefferson's argument that blacks and whites were morally equal was based on his conviction that blacks had sufficient intelligence to reason and make choices. All humans, regardless of race, are moral agents and created equal in this respect. This, and the corollary that slavery was immoral, seems never to have been doubted. Rather at issue was the degree to which the black capacity to reason was the equal of the white. It was this intellectual standing of the races that troubled him. This is a question that is not unusual for the time. Slaves were not encouraged to develop their intellectual gifts, and so the "evidence" was bound to suggest there were differences, and Jefferson realized this. The point is that he approached the question more or less from an "experimental" and "empirical" stance. For him the question was: Were blacks different, and perhaps intellectually inferior, by nature, or were differences in intellectual achievements due to circumstance? (Now, other contemporaries of his had no doubts on this score. Benjamin Rush and Samuel Stanhope, for example, believed that blacks were intellectually equal to whites. Still, Jefferson asked the question, and sought an answer.) The evidence on the surface supported inferiority, but he wanted to see if it could be disconfirmed. He considered two senses in which the environment may explain the apparent "inferiority" of black faculties. First, blacks may act the way they do because of the influence of the African physical environment. If their environment were to change—if they were to be placed in a new one for long enough—so too might their behavior. Second, enslaving them in plantation culture places them in a social environment that may stunt their intellectual growth. They seem inferior because they have not had the opportunities whites have had. These then were grounds for caution: "the opinion that they [the black] are inferior in the faculties of reason and imagination, must be hazarded with great diffidence."[22] It was only then an opinion or a "suspicion,"[23] or, in modern parlance, a hypothesis: this led him to seek out evidence to the contrary.

Shortly after returning home from France in 1791, he learned about the talented black mathematician Benjamin Banneker. After examining some of his work, Jefferson wrote to Banneker: "Nobody wishes more than I do to see such proofs as you exhibit, that nature has given to our black brethren, talents equal to those of the other colors of men, and that the appearance of want of them is owing merely to the degraded condition of their existence, both in Africa and America. I can add with truth that nobody wishes more ardently to see a good system commenced for raising the condition both of their body and mind to what it ought to be, as fast as the imbecility of their present existence . . . will admit."[24] He then sent Banneker's work to the Marquis de Condorcet, saying he was "happy" to provide evidence that a Negro could become "a very respectable mathematician. . . . I have seen very elegant solutions of geometrical problems by him. . . . I shall be delighted to see these instances of moral eminence so multiplied as to prove that the want of talent observed in them is . . . the effect of their degraded condition, and not preceding from difference in the structure of their parts on which intellect depends."[25] True, he uses the word moral, rather than intellectual, which leads Winthrop Jordan to see this as Jefferson hedging his bets,[26] for Jefferson had already concluded that blacks were fully morally developed, and the thought experiments were to determine the intellectual equality.[27] Still he was nevertheless making a reasonable effort.

One can argue that a truly open-minded person would have accepted examples of intellectual giftedness such as that exhibited by Banneker as strong evidence of the intellectual equality of the races, especially given the "degraded" and disadvantaged conditions under which Banneker must have learned and worked. Certainly, Jefferson could have pointed to the abundant cases of many privileged but mentally challenged whites to convince him that the intellectual bar for comparison was not even very high. Still, he did not accept this or other evidence as disconfirming the hypothesis, which was that blacks were likely mentally inferior to some degree, though again, he was always keeping an "open" mind; and all the while he never doubted that blacks were morally equivalent to whites and so deserved to be free. In regard to mental equality and to mixing of races, he could not completely overcome his own biases, and so also could not see clearly enough how the geographical conditions of slavery created overwhelmingly powerful forces stunting human potential and producing predictable and unfairly biased outcomes. He tried,—and he tried in public—but not hard enough. Still his conclusions troubled him. And again, these conclusions were about the equality of intellectual capacities, and not the fact that blacks had the intellectual capacities needed for moral equality, and so they were not about the evils of slavery. This is why I believe that on the big moral questions Jefferson was not using compartmentalization as a form of self-deception. At this level he was not blinding himself to the evidence, nor narrowing his moral commitment. Rather it was the failure of a well-intentioned person to try hard enough on the specifics and details. These are examples of small self-deceptions and resulting moral drift. If this happens to Thomas Jefferson, it is certainly understandable how it happens to the multitude of less capable individuals and also how for them, the details can obscure the bigger picture. They would find it comfortable not to question the results of compartmentalization, and to limit their moral sense of what is right to a narrow, self-interested, and instrumental range. And if the rest of the society has already narrowed its gaze, then the moral audience contracts, moral drift sets in, and it becomes increasingly difficult to think oneself out of the situation.

Clement Eaton[28] stresses this point when he argues that while during the period of the American Revolution, when the great ideas of human freedom and equality were expanding everyone's view, virtually none of the "great Virginians" [i.e., Washington, Jefferson, Madison, Patrick Henry] ... failed to put himself on record as favoring emancipation."[29] These were children of the Enlightenment, and they were playing on the world stage. But by the middle of the 19th century, attitudes of Southern plantation owners had narrowed. The South became inward and insular. Instead of looking forward in terms of progress and breaking new social and intellectual ground, Southerners came to look backward, romanticizing aristocratic and feudal Europe and believing that they were holding on to those chivalrous ideas of honor, nobility, and noblesse oblige.[30] "Southerners of the upper class found a certain satisfaction in comparing their civilization based on black dependents, with medieval manors, knights on caracoling horses, and humble serfs. Slavery nourished a quixotic pride in the ruling class and it deflected southern literature from dealing realistically with southern problems."[31] Between 1790 and 1850 a striking change had taken place in the quality of the leaders produced by the South. "The Southern plantations of the eighteenth century had bred such liberals as George Mason, James Madison, George Washington, Thomas Jefferson, Charles Carroll, James Monroe, and John Randolph . . . [In] 1850 their places were occupied by men of smaller stature . . . They were warped by strong sectional prejudices, they lacked a catholic point of view, and they had lost the magic glow of republicanism [and] . . . their breadth of vision was limited by their vested interest in slavery and a devotion to the romantic."[32] When attacked by abolitionists, their defense was to continue to narrow their vision. They enacted a host of laws and regulations which restricted access to information and freedom of speech and the press, and they censored instructional materials in schools and universities. Compared to the propaganda and censorship of the 20th-century dictatorships, this was mild; still it contributed enormously to a narrowed vision and to the problem of having these habits and practices reinforced geographically. Everyone was doing it, so even if it was not right, it was at least normal.

Slaveholders of the 1830s and 1840s had, by and large, convinced themselves that the institution of slavery was good. "Slavery may have been immoral to the world at large, but to these men, notwithstanding their doubts and inner conflicts, it increasingly came to be seen as the very foundation of a proper social order and therefore as the scene of morality in human relationships."[33] Southern pride no longer took account of contributing to and being a part of humanity at large, or even of the Union, but rather pride and allegiance resided in the provincial boundaries of the plantation: "The plantation [was] . . . raised to a political principle."[34]

By the eve of the Civil War, most upper-class Southerners could not or did not try to critically examine the conditions of slavery. They accepted the outcomes of their compartmentalized world as though they were necessary, just, and natural. By accepting a narrow view of what ought to be, a slaveholder could readily be convinced that since he did not create slavery, or "mistreat" his slaves, he then was not acting immorally. Indeed, he may even have seen himself as a decent person, doing the best he could in a less than perfect world; or even as a good person, providing guidance and protection to an "inferior" group; or as a person committed to great and noble cause, as stated by the vice president of the Confederacy, when he proclaimed the Confederate states to be the first in history to be founded on the moral principles of slavery.

Not all of Jefferson's contemporaries subjected the system to such scrutiny, and as we noted, by the middle of the 19th century hardly any slaveholders did, though they were confronted by the irrefutable evidence that blacks were moral agents and intelligent human beings. The evidence was there in their daily encounters with blacks, if only they could see through the fact that the rules of place limited black capacities, and that even the need for rules such as those prohibiting literacy and holding blacks morally responsible for violating rules, already assumed the existence of these very intellectual capacities the slaveholders were not willing to admit existed. So, to hold unquestioningly, as did the supporters of slavery, that blacks are not fully human (and that slavery at best is a paternalistic system that helps guide such "inferior" people) is an act of self-deception.

As [Mordekhai] Thomas puts it: "Any attempt to deny the humanity of a group requires significant and sustained . . . maneuvering . . . . Any human beings who deny the humanity of others with whom they regularly interact are engaged in the attempt to live as a lie what is revealed to them as manifestly false by their most fundamental ongoing human existence."[35] Self-deception, and of course moral drift, were clearly the case in the antebellum South. Their existence, though based on decisions by individuals, led these same individuals to not be able to see what was really happening. It allowed individuals to create and support a horrifyingly evil system, and yet not be aware of the fact that they were doing evil. This is the power of self-deception, and it supports the theory's argument that people do not do evil willingly.

Nazi Germany and Hitler

The structure of 20th century industrialized Germany and its use of compartmentalization were much more complex than in the American South. The differences are evident as soon as we think of the relationship between victim and victimizer. In the case of the antebellum South, compartmentalization did not necessarily mean that blacks were out of the view of Southern plantation owners (even though the day-to-day operation of the plantation was often in the hands of overseers and drivers), or that these owners did not often see the more horrifying consequences of their actions. Rather compartmentalization (among other things) limited blacks' opportunities and created and reproduced stereotypes that seemed in the eyes of slave owners to justify the foundations of slavery. Compartmentalization helped create inferiority, which then was used as evidence to justify compartmentalization. In Nazi Germany, we find compartmentalization to be used in this way too, but we also find it to be used more frequently than in the American South to actually remove evil and its effects from view, and even to keep some of it secret. I will consider the effects of compartmentalization in three steps. First are the ordinary Germans who lived during this period and went along with Nazi policies but may not have known the details; second are those who executed the orders (e.g., Eichmann, Höss, and the SS prison guards); and third are those who initiated the evil (especially Hitler).

Ordinary Germans   When the party came to power it initiated a set of rules that, as we have seen, began to compartmentalize and narrow the world of the Jews, and also that of the Germans, leading the latter in the direction of moral drift. The average German in this restricted world supported Hitler—perhaps not everything he did, but much of what he did. Individuals who support dictatorships purposefully give over their own right to decide; they no longer control their own critical views and expectations. What is right becomes determined for them by those in authority. The Nazi vision of what was right was to encourage hate and conflict with those who were not part of the *volk.* Hate, more than any other emotion, narrows our moral vision. We do not want to know or see any other evidence than what fuels hate. The day-to-day expectations of being good are narrowed—be good to the Aryan neighbor, to the ideal of the state and to the purity of the race, but not to any "*other.*" Even if one did not take this seriously, it became increasingly difficult to do anything about it. And so, an average German could assume a view of decency, which is that "I have not hurt anyone directly, I have not turned anyone in, I have not been malicious," and use these minimalist precepts as evidence that he is emotionally resisting Nazism, without doing anything outward and public.

Unfortunately there were many Germans who did adhere to much of Nazi ideology. Eric Johnson has shown that the Gestapo did not have to supervise the German people closely because not only did they conform to Nazi policy, they actually abetted the Nazis. "Because average citizens were so often willing to keep watch over and denounce fellow citizens whenever they stepped out of line, . . . relatively few secret police officers were needed to control a German population that was quite ready and able to control itself."[36] So many Germans could be relied on to turn Jews in to the authorities that, as we have seen, the Gestapo

Chief Heydrich could count on them to be the eyes and ears of the police, and so as not to have to jump prematurely into a ghettoization of Jews. And as the Jews became isolated, the German citizens did not have to probe deeply about what took place after the they were gone, for it was no longer their responsibility: out of sight, out of mind.

That there were ghettos and concentration camps was common knowledge. But the actual extermination of the Jews was supposed to be a secret. How widely known this secret was is a matter of debate, but this is not terribly important here, for geographical segmentation meant that if one does not want to know, one can block it out. And if one hates a group, one certainly would not seek out evidence that would change one's view—and if one encountered it anyway, self-deception would eliminate its effects. One way the evidence, even if encountered, could be "discounted," is that the geographical conditions of concentration and ghettoization (as with the slave plantations) would "produce" a victim that ever more conformed to stereotype. If the Jews are supposed to be clannish, then forcing them together makes them more so. If the Jews are supposed to be unclean, then life amidst the squalor of the ghetto and the concentration camp will make this come true. So, even if one became aware, it could be used as evidence to reinforce existing prejudices. As Arendt perhaps overly states it: "German society of eighty million people had been shielded against reality and factuality . . . by self-deception . . . the practice of self-deception had become so common, almost . . . a prerequisite for survival."[37]

What percentage of the population was involved in this moral drift and self-deception? It is impossible to say. Some have argued that it involved most Germans in some form.[38] But even so, there were significant exceptions. The most dramatic are those Germans who harbored and protected Jews. Their resistance was mostly secret, but heroic, and the numbers who helped were significant. These individuals were able to see through the use of compartmentalization that helped create hate, moral drift and self-deception.[39] It was possible to be good in Nazi Germany. But self-deception and moral drift were so deeply a part of the culture that to be good required acts of heroism.

Those who executed orders What role did compartmentalization and self-deception play in the lives of those who were directly involved in executing Nazi orders? Here the relationship between compartmentalization and self-deception can be seen most deeply in the context of segmenting decisions into steps and stages along a chain of command. Tzvetan Todorov, in *Facing the Extreme: Moral Life in the Concentra-*

tion Camps, provides an insightful example of how individuals in their daily routines can be enmeshed in small tasks done well and with dedication that unfortunately can contribute to a vast project of evil without creating in these individuals the slightest feelings of responsibility. He asks us to consider that at one end "of the chain are people like Reinhard Heydrich: his sleep is never disturbed by the millions of deaths that took place on his orders. He never sees a single suffering face; all he does is manipulate large and odorless numbers. Then come the policeman, a Frenchman, let's say. His carefully circumscribed job is to ferret out Jewish children and take them to a camp where they are turned over to German personnel. This policeman does not kill anyone; he merely carries out a routine order to arrest and expedite. Now Eichmann enters the picture, his purely technical job consisting of making sure that a certain train leaves Drancy on the fifteenth and arrives at Auschwitz on the twenty-second. Just where is the crime in that? And then we have Höss, who gives the orders to empty the trains and lead the children toward the gas chambers. And finally, the last link: a group of inmates, a special commando that pushes the victims into the gas chambers and releases the lethal gas. The members of this commando are the only people who kill with their own hands— although they quite obviously are victims themselves, not executioners. No element in the chain (which is actually much longer) feels responsible for what has been accomplished; the compartmentalization of the work has suspended considerations of conscience. Only at either end of the chain is the situation slightly different."[40]

Those on one extreme end—the ones who actually push the levers— the Kapos—are not really of interest here because they hardly were involved in the decision and barely had the will to say no. Doing what they did allowed them to live, at least a while longer. They too were victims, as Todorov says. But those along the rest of the chain had more options. The guards running the camps did not have to be there. They could have asked for, and would have been granted, transfers. Transfers would have hurt their careers, and possibly endangered their lives in that they may have been assigned to front-line combat responsibilities, but they could have resisted. According to the Nüremberg documents, "It was surprisingly easy for members of the extermination squads to quit their jobs without serious consequences for themselves . . . not a single case could be traced in which an SS member has suffered the death penalty because of refusal to take part in an execution";[41] the SS testimony supports this: "as far as I know, no serious consequences resulted from [asking to be transferred in order not to participate in the mass murders]. However, it was clear that in general such people could not count on promotion for the foreseeable future. . . . I know of no instance where apart from a block on promotion, or a punitive transfer, a unit leader was sent to a concentration camp or sentenced to death. Concerning the lesser ranks, I also know of no instance in which refusal to take part in the shooting of Jews resulted in anyone being sent to a concentration camp or being sentenced to death."[42]

Even those who were employed in the camps did not have to be excessively brutal in the execution of their "duties." By far the majority were not psychopaths and sadists: "nothing could be more mistaken than to see the SS as a sadistic horde driven to abuse and torture thousands of human beings by instinct, passion, or some thirst for pleasure. Those who acted in this way were a small minority . . . the predominant type was . . . a conformist."[43] Conformists gladly relinquish the questioning of practices for the comfort of acceptance. Conformists are prone to moral drift and self-deception. As Arthur Seyss-Inquart, the Nazi governor of Austria remarked: "There is a limit to the number of people you can kill out of hatred or a lust for slaughter. But there is no limit to the number you can kill in the cold, systematic manner of the military 'categorical imperative.'"[44] These guards were like other recruits. Many of them came from military units "with hardly more crimes in its record than any ordinary unit of the German Army . . ."[45]

So how could they do what they did? Were they doing evil willingly? They were certainly doing evil, but they may not have seen their own actions as such because of the roles of compartmentalization, moral drift, and self-deception. They were part of an anti-Semitic culture and the government's political rhetoric was one of hate. Commitment to critical thinking was already relinquished as soon as they gave their support to a totalitarian state. In the Nazi case, among the most important demands made of its followers was the complete immersion of the individual within the mind and soul of a racially pure group so that, according to Hitler, for the ordinary German, belonging to the volk would make even the nation appear to be nothing and insignificant, and the meaning of belonging to the more specialized units, such as the SS, according to Himmler, would require of an individual "total sacrifice of his person to the accomplishment of his duty . . ."[46] This ideology presented the Nazi mission in terms of moral rectitude—by belonging to the volk and following Nazi principles, the Germans as a whole were doing the right thing, and it encouraged a particular individual to see himself as obediently following orders, as simply a cog in a bureaucratic machine. To make this view plausible even for those involved in the final solution, the actual killing was made as impersonal and antiseptic as possible.

Nazism provided for its followers a world turned upside down. Even so, there were difficulties. Though they mightily resisted, SS members found themselves feeling pity for the victims. These feelings had to be suppressed. Eicke said of pity: "Pity for these men would be unworthy of an SS man: in our ranks there is no room for softies, they are better off entering a monastery, we need men who are tough and committed . . . and SS men . . . must be capable of annihilating even his closest family if they rebel against the state or the ideas of Adolf Hitler."[47] But even hardened SS troops would succumb to doubt when the killing became too personal. Before the concentration camps and relative impersonality of the gas chambers, some of the members of the mobile killing units suffered remorse: "After the bloodiest actions [-] several SS men killed themselves, and others got methodically drunk."[48] Even when the killing took place in the gas chambers, the world of the concentration camp—its geographical layout—was organized as much as possible to compartmentalize responsibility and to make the life of the SS as normal as possible. This coupled with all of the other aspects of self-deception, moral drift, and compartmentalization of Nazi life help us understand, but by no means excuse how otherwise sane and normal individuals were able to commit such atrocities. It shows how individuals simply did not want to know, and would not make the effort to know, that what they were doing was wrong. Moral drift and compartmentalization are not a justification, but they are part of an explanation for how people can have free will and do evil, because they do not see the good. True, this argument, as we have noted before, makes evil banal in that it is no longer something out of the ordinary. But that may well be what evil is. If it were only monsters who committed evil, then we would not look for it in ourselves, but only in others.

What of the individuals higher up the chain of command? We find the same thing here. But because they were giving orders, they could increase the distance between themselves and their victims. This allowed them to continue to believe that they were simply cogs in a bureaucratic wheel, obeying orders, and not personally doing harm to anyone. The quality of geographical compartmentalization abetted such thinking, especially when places became thinned and highly specialized, with a bureaucratic and antiseptic quality. Being in a distant office and writing memos provided the illusion of apartness from the actual deed. Running a railroad was less personal than guarding prisoners. Closer up to the act, the gas chamber was far more impersonal and antiseptic than a knife or a bullet. Language too helped compartmentalize. The extermination of Jews was always discussed in code or "language rules" as the Nazis called them. Even the concept of extermination was called the "Final Solution," "evacuation," "special treatment," or "deportation."[49] The identity of individuals became erased—they were assigned numbers and were described as "cargo" or "items" or as nothings, as in "ninety-seven thousand have been dealt with."[50] According to Arendt, "For whatever other reasons the language rules may have been devised they proved of enormous help in the maintenance of order and sanity in the various widely diverse services whose cooperation was essential in the matter. Moreover, the very term 'language rule' was itself a code name; it meant what in ordinary language would be called a lie ... The net effect of this language system was not to keep these people ignorant of what they were doing, but to prevent them from equating it with their old "normal" knowledge of murder and lies."[51] The net effect was self-deception.

The types of people recruited into the party and to its higher echelons were particularly susceptible to self-deception. Höss, whom we have already encountered as the commandant of Auschwitz, and typical of the Nazi official, was a model of the self-deceived individual. In his autobiography, which he wrote between the time of his sentencing in the Nüremberg trials and his execution, we encounter an individual who even at the end of his life was not able to see that what he did was wrong. He continued to believe he was only obeying orders, and not personally responsible for harming anyone: the archetypical conformist. He tells the reader that he never wanted to inflict pain and suffering, not even on animals. He was raised to obey—"to be absolutely obedient to the point of the most painstaking neatness and cleanliness,"[52] and that he continued to obey absolutely as a soldier, for that is what soldiers must do.[53] But his was not only obedience to authority, it was also a conviction that what the authorities believed was right and just. "Whatever the Führer or Himmler ordered was always right."[54] "As an old-time member of the Nazi Party, I believed in the need for the concentration camps."[55]

He could not escape observing the killings, which caused in him some squeamishness. But these basic responses to human suffering were overridden by his sense of blind duty and belief in the goodness of the cause, his need to conform, and by seeing that the executions were clean and bloodless.[56] Still he was not completely at ease, if not because of his conscience, then because of some basic sense of animal sympathy that still tugged at him. He claims that others too felt it, even the higher officials who left Berlin headquarters to inspect the killing operations. "All of them were deeply impressed by what they saw. Some of them who had lectured very fanatically about the necessity of this extermination became completely silent while viewing the "Final Solution of the Jewish Question" and remained so. I was asked repeatedly how I and my men could watch these proceedings day after day. How could I stand it? I gave the same answer time and time again, that only iron determination could carry out Hitler's orders and this could only be achieved by stifling all human emotion. Even [SS General of the Gestapo] Mildner and Eichmann, who had a reputation for being truly hard, said they would not want to change places with me. No one envied my job."[57]

Observing others suffer did not soften the determination of these leaders. On the contrary. In their perverse way, it seemed to convince them even more that they must continue the process. Höss describes how, at a dinner after such an inspection, when the wine flowed freely and true feelings showed, Eichmann's revealed even greater obsessions to do the job completely: "of destroying every Jew he could get his hand on. Ice cold and without mercy, we had to carry out this annihilation as quickly as possible. Any compromise, even the smallest, would bitterly avenge itself later on."[58] This grim determination of his superior then reinforced Höss's resolve. "I have to confess openly that after such conversations with Eichmann these human emotions seemed almost like treason against the Führer."[59]

After the war was over, after he was taken into custody, and even after he was sentenced to death for war crimes, Höss did not see things differently. He claims that he was sorry for the extermination of the Jews. But why? He still believed Jews were evil and deserved to die. The reason he is sorry is that the extermination was not good instrumentally. It did not produce the desired effects: it was absolutely wrong because it created world enmity against the Germans and sympathy for the Jews.[60] And what of his personal role in the killings? Even there, he was only an instrument, and then a clean and guiltless one: "I never personally mistreated a prisoner, or even killed one. I have also never tolerated mistreatment on the part of my subordinates."[61]

Höss mentions Eichmann (an important SS official organizing the exterminations) and his speech rallying those around him to do the right thing. For Eichmann, as for all of those in command, the right thing was absolute obedience. Eichmann identified so completely with the system that on the official date of German surrender it dawned on him that he would no longer be part of this or any other process: "I sensed I would have to live a leaderless and difficult individual life, I would receive no directives from anybody, nor orders and commands would any longer be issued to me, no pertinent ordinances would be there to consult—in brief, a life never known before lay before me."[62] Eichmann claimed he did not hate. He was simply doing his duty. His degree of compartmentalization and lack of awareness is almost a parody. About his visit to Chelmno where victims were still being gassed in trucks, he says: "I did not stay to watch the whole maneuver. I couldn't stand the screams: I was far too anxious . . ." As for his visit to Auschwitz: "I preferred not to watch the way they asphyxiated people . . .they burned the corpses on a gigantic iron grill . . . I couldn't stand it: I was overcome with nausea." In Treblinka, "I stood off to one side; I would have liked not to see anything at all." And when he ordered Hungarian Jews to march to Vienna: "I didn't see them myself; on principle, I refused to watch these oppressive sights unless formally ordered to do so."[63] Still he believed he was innocent of responsibility.

A most important part of his argument was that he could find no one else in his position who did not do the same thing. There was no one among his peers who was opposed to the Final Solution and who could through disobedience set a counterexample. But this too was another example of self-deception, for he mentions elsewhere a case in Hungary[64] and he may well have known about the disobedience of the SS in Denmark.[65] These however did not register.

The entire concept of toughness in the face of suffering, of having to "bear the burden" of inflicting pain for a better and higher set of ideals, was one gigantic form of self-deception, that we find throughout the Nazi ranks. Speer, who was a close confidant of Hitler, and perhaps the one among the top who had some sense of guilt and awareness, describes how he compartmentalized his relationship: "I felt myself to be Hitler's architect. Political events did not concern me. . . . I felt there was no need for me to take any political position at all. Nazi education, furthermore, aimed at separatist thinking; I was expected to confine myself to the job of building." His compartmentalization did not change when his position changed from architect to Minister of Armaments: "The task I have to fulfill is an unpolitical one."[66]

According to Arendt, the single most important factor in creating evil was the thoughtlessness associated with self-deception. She says of Eichmann, whom she saw as an important, but representative figure in the process: "it was sheer thoughtlessness—something by no means identical with stupidity—that predisposed him to become one of the greatest criminals of the period . . . such remoteness from reality and such thoughtlessness can wreak more havoc than all of the evil instincts taken together."[67] Todorov agrees, saying "compartmentalization and the bureaucratic specialization to which it gives rise are at the root of this absence of feelings of responsibility one finds in those who carried out the Final Solution, as well as in every other agent of the totalitarian state.[68] Compartmentalization by itself is neither bad or good, but in the hands of totalitarians it almost always is used to produce evil."[69]

Hitler: doing evil willingly?   What of the highest leaders and of Hitler especially? Again, at the risk of making someone who may well be a monster (in the sense of not having a choice, being hardwired to do evil, or being criminally insane and not responsible for his actions) into someone who is more like the rest of us, I believe that the evidence again supports self-deception. He did not know the meaning of what he did. Some of his blindness was due to rage, but most of it was due to the narrowing of his world through fantasy and hate that did not brook contradictions. His awareness was intense but extremely narrow and distorted, for his mind was closed.

Hitler was blind to evidence and even to open-mindedness. A strong undercurrent in *Mein Kampf* is that an open mind, and a world that encourages this—a world of questioning, in which all individuals have the right to search for truth, and in which each individual is to be respected and protected by principles of democracy and equality—are themselves largely products of the Jewish conspiracy, for they benefit the Jews. These principles subvert the real truth about humans. This truth is that human beings are inescapably molded by and products of volk, blood, soil, and are enmeshed in a Social Darwinian struggle, and that the German volk are the strongest and purest and destined to rule. The so-called natural rights of life and liberty are falsehoods. The only true right is that of racial purity.[70] This he believes is the nature of reality, to which many are blinded because of the deceptive doctrine of natural rights that is promulgated by the Jews. Because of them we cannot learn from nature. We become too sentimental and try to protect those who are weak and cannot survive without assistance. Because of them "man is hysterically concerned that once a being is born, it should be preserved at any price."[71] It is in the interest of Jews to propagate such deceptions because they do not have a place in nature, and thus need to subvert it with an ideology of equality. This is why, when he realized he was an anti-Semite he also "ceased to be a weak-kneed cosmopolite."[72] Jews were responsible for perpetrating the myth of democratic and egalitarian principles—the Jewish democratic idea as he called it.[73] The Jews promulgate the ideas of universality and equality to dilute the power of real people and cultures, and thereby run counter to nature. They are unable to maintain a territory and government of their own, and so must live, like parasites, off of the bodies of other nations.[74]

Open-mindedness and free access to knowledge actually prevent people from seeing clearly. Instead of open-mindedness there must be a conviction that is so extreme and intense that it turns into fanaticism, rage, and hatred. In 1923 in Munich, Hitler claims that "For the liberation of the people more is needed than an economic policy, more than industry: if a people is to become free, it needs pride and willpower, defiance, hate, hate and once again hate."[75] Hatred was the impulse behind his movement. Hatred unites. And hatred, along with fanaticism to the cause, are supreme virtues. He states the same thing in *Mein Kampf.* "Hate is more enduring than aversion, and the impetus to the mightiest upheavals on this earth has at all times consisted less in a scientific knowledge dominating the masses than in a fanaticism which inspired them and sometimes in a hysteria which drove them forward . . ."[76] The greatness of social movements lies in "fanaticism and intolerance . . . fanatically convinced of its own right, it intolerantly imposes its will against all others."[77]

Hate, ferocity, fanaticism, intolerance, and vengeance were all virtues in *Mein Kampf.* According to Lukacs, hate was a primary force in Hitler's character—"hate and his consciousness of those hatreds."[78] The bitter and brutal struggles of his Social Darwinism were, to use Lukacs' word, "fortified,"[79] by the strong emotion of hate. . . . His lesson was "to hate and to be hard . . . a lesson devoid of love." According to Goebbles, Hitler kept telling him he "had learned to hate." Hate, in the hate against those who hated Germans, for him was "God's most beautiful gift bestowed on us."[80]

To have fanaticism and hatred close our minds and narrow our visions is, for Hitler, a good, and it must be supported by state instruments such as propaganda and censorship. Even science must be secret and occult: "Only when knowledge assumes once again the character of a secret science, and is not [the property] of everyone, will it assume once more its usual function, namely, as an instrument of domination, of human nature as well as that which stands outside him."[81]

A strong case can be made then from these and other remarks that Hitler believed what he was doing was righteous; and that his was a case of extreme and immensely tragic self-deception. Certainly, many eminent historians like Trevor Roper have claimed that is the case.[82] It seems that Hitler is deceiving himself on two levels: he deludes himself about the facts concerning the Jews. They are not a race (and nor are the Germans or Aryans) and the Jews simply are not doing what he

and other anti-Semites claim they do. But perhaps even more importantly, he seems to have deluded himself into thinking that the remedy
for self-deception—the edifying qualities of an open and public search
for evidence and alternative viewpoints that are found in democratic
and cosmopolitan society—is itself a Jewish plot. So there is really no
way out of this prejudice. He is blind—he has blinded himself—both
to the truth, and to the method of finding it. Everything then becomes
twisted.

Again, compartmentalization is linked to the process of self-deception, even for Hitler. He will not tolerate contradiction. He surrounds
himself with sycophants and yes-men. He also uses coded language to
discuss his plans, and thus does not have to be reminded of what it
really means, and he does not want to literally see or sense the results of
the slaughter. This applies not only to the concentration camp victims,
but also to the ravages of war inflicted upon Germans themselves. According to Lukacs, "there was evident unwillingness to visit or even to
look at photographs of severely bombed German cities, or at pictures
of the German refugees trekking westward in the brutal winter of
1944–1945, pictures that he, at least on occasion, pushed aside. . . .
Here we are faced, again, with a duality of character. . . . He took no
sensual pleasure in the sufferings of his enemies or victims . . ."[83] And
those close to him claimed that he was kind to his friends and to animals.

It must be understood that self-deception is not an excuse for doing
evil. It simply is a means of helping us understand how evil can be
done while having us retain both the assumption of free will and the
idea that the good is compelling. While I want to believe that self-deception is an explanation for even the most horrible atrocities and that
the evidence above supports self-deception, I must also admit that the
very same evidence can be used to support the opposite claim—that
Hitler knew full-well that what he was doing was evil, and that he embraced it willingly, even joyously. Hitler then was doing evil willingly
and his justifications—that what he was doing was good—were simply
rationalizations, for most other Germans were simply not prepared to
do the same.[84] He kept the actual implementation of the final solution
a secret (though not his hatred and annihilationist remarks) because
he knew Germans would see that it was wrong. Even the fact that he
did not want to see what was taking place should not be used as evidence of self-deception, but rather simply as a way of focusing his attention on—or of not being distracted from—his bigger and more evil
vision. In other words, it was not an attempt to keep evidence of "not-
p" from overwhelming his belief in "p" but rather a decision that allowed

him to realize that the act of knowing more about "not-p" would simply take up too much of his time and energy.

We can read the evidence this way, but to do so would not only require that we abandon the claim that we do not do evil willingly (and that the good is compelling), but that for people like him, evil is more attractive than good. His actions then stand out not only because of their enormous impact, but also because of the willingness and enthusiasm with which he embraced it. Again this may be an accurate way of dealing with him and others who some think of as moral monsters. But it means we must then abandon the view that the good is real and compelling, and argue instead that good and evil are relative. If, however, we think that this is giving away too much—if we do not think it is true that good is not as attractive as evil and if we believe that people do make choices—there seems to be no alternative but to come back to the arguments of the theory and explain evil as a lack of will and a matter of self-deception. These arguments in turn place Hitler's acts as close to, if not at, the end point of a continuum along which any one of us can move with tragic consequences if we are not careful and if we do not exercise the will to resist.

MOVING BACK TO THE GOOD

We began this chapter with the understanding that place and its capacity to compartmentalize is a double-edged sword. It can help focus our attention so that we can then see more clearly and create variety and complexity and impart this to the world, or it can help to cloud our view, narrow our horizons and sense of responsibility, and pull us down the path of moral drift and self-deception. If this has happened for a society as a whole, then how can it be turned around?

A starting point is this: although hegemony and uniformity are goals of evil, no regime has been completely successful in erasing differences. As we saw, despite the density of power that spread over the concentration camp, cracks and crevices emerged in its landscape that allowed for sites of resistance. What could happen in a place like this would certainly happen in a larger one. Even when people are motivated by similar objectives of self-interest, differences of opinion and viewpoints arise, and while these may not be good ones, simply the fact of difference opens up the possibility of greater awareness that can lead to the good—though to be good, to speak the truth, and to help others when surrounded by a large number of citizens who have relinquished their duty to make choices, is itself a heroic act. Fortunately, in the cases we discussed, there were heroes. These were people who saw clearly. They risked their lives to impart this truth as widely as possible[85] and/or by translating this truth into acts of kindness, care, and justice. It can be expected that the more these efforts to transmit truth and provide care are made visible to the rest of the closed society, minds would open and the tide of moral drift would turn. This is so because the good is compelling and making it open and public would force even those in charge to be less self-deceived.

To be reassured that this is the case, we need only remind ourselves that these bad societies banned freedom of expression and reduced the variety of practices precisely because they knew that appeal to reason and to varied points of view help pry individuals loose from the prejudices of situated and self-interested positions. Even so, we must be realistic about how slow, halting, and dangerous the results of making the good visible would be in a closed and hateful society. Good acts may initially not be seen as such. Sites of resistance may be quashed, if, all at once, they became open and transparent and showed the rest of the society exactly how a human being should behave. On the other hand, if all acts of goodness in an evil society were hidden from view, these acts would succeed in keeping some victims alive, but would not serve to counter the trend of self-deception and evil. The issue is one of degree. In evil societies, even good places must be secretive to a considerable extent, but to succeed in countering evil, they must also strive to be as open as possible.

An example of a good place in an evil society that pushed the bounds of openness as far as possible is the French village of Le Chambon that helped save approximately 5,000 refugees from the Nazis. It will be recalled from Chapter 4 that this French Huguenot settlement had a longstanding tradition of pacifism and operated on the principle that no one in need should ever be turned away. The village never denied it offered refuge to the Holocaust victims. Its reputation as a haven may even have softened Nazi aggression against it. The Nazis never killed the villagers and, mysteriously, phone calls would be made from some unknown person alerting the villagers of each raid. This gave them enough time to further sequester the refugees.

It is not unreasonable to assume that this degree of openness shamed the local Nazi authorities, and blunted their hate. Still, Le Chambon needed to practice secrecy as well, for it remained in danger. The villagers organized themselves into cells so that no one knew the whole picture, and when questioned by the authorities, they never revealed the names of the refugees or where they were kept. Both the boundaries and the surfaces of these places were still disguised. Most tellingly, the reputation of the community for kindness did not prevent an assassination threat on its leader, who himself had to go into hiding.

The point is that these and other forms of secrecy still needed to be practiced. Important parts of Le Chambon's efforts at nonviolent resistance and pacifism had to be disguised. Still, its degree of openness allowed it to have more influence than would have been the case if its good deeds were done in complete secrecy. If there were more examples like Le Chambon, the tide would have turned more quickly. Indeed, Nazi virulence seemed to lessen in a country like Denmark, which as a nation, announced its opposition to the principles of anti-Semitism.

The openness of Le Chambon moved the surrounding areas a small degree away from self-deception. For Le Chambon to be still more open and yet survive would require that it be in yet a better society, one that had become so by more examples of openness. How long such an iterative process would take to significantly improve a Nazi Germany, or a Stalin's Russia, or an antebellum South is impossible to tell. It is not impossible to say however that when they were at their worst, these societies would not have tolerated the immediate introduction of larger-scale and more open examples of passive and nonviolent resistance such as was preached and practiced by Gandhi and Nehru in India and by the Civil Rights demonstrations of the 1960s in the U.S. These were possible and successful because there was already a core of sympathy among the other citizens to the principles of these movements. Indeed, the American civil rights movement with its nonviolent geographic strategies of sit-ins and pickets, succeeded in large measure because it embarrassed white Americans by confronting them with the fact that their practices were not meeting their own professed visions of America. The same can be said of the successes of nonviolent protests of Gandhi and Nehru in India. They showed the British that their colonial behavior was ultimately hypocritical. These movements, and others such as the liberation movement in the Philippines, and the student protest movement of China and its Tiananmen Square, were successful also because a degree of openness in communication allowed them to take their cases to a cooler and more distant moral public—a world audience—uncluttered by the passions and prejudices of the local.[86]

These sites of resistance are examples of how individuals create places that increase our awareness and increase variety and complexity. Creating such places leads to moral progress. Those who are morally gifted seem to know this without the guidance of a moral theory. They simply have an intimation of the good. Others need a guide. In either case, if places are created according to intrinsic judgments, moral progress can be made, though it is likely to be halting and painfully slow. This path to moral progress emphasizes how the good acts as a lure allowing the instrumental to become like the intrinsic.

But what happens without an articulated moral theory as a guide? I want to show that a rougher sense of direction can be drawn simply from our engagement in instrumental projects themselves. Specifically it can be encountered in the experience of mastering a project and doing the project beautifully. Doing something excellently and beautifully introduces us to the idea of contributing to a purpose larger than ourselves. By no means is doing something well and beautifully an unerring guide to the good, but it does possess qualities that are akin to those in intrinsic judgments. As Kant would say, it serves as a propaedeutic to the good. Let us consider how projects that are excellently and aesthetically done can make us more sensitive to the good.

Excellence as a Bridge to the Good

Mastering a task, doing it well, and especially excellently, is an important part of instrumental activity. It also introduces us to the idea of a goal, of progress, and of contributing something that can be of help to others. Consider a child who learns to read. If this is not her explicit goal, it is certainly a goal society assigns to her; and the child can make progress toward it. She can progress slowly or rapidly, and her progress can be compared to other children who are attempting to master this task. In this case, instrumental progress is acquiring a skill: a child learning a language or a game of chess, a baseball player becoming a better batter, a violinist becoming a virtuoso, a man becoming a better parent. If the skill is acquired well or even excellently, the accomplishments can and will stand as a model. It will be worth emulating. Her mastery helps her and others.

The skills we acquire involve using geography. Place-making not only undergirds projects, but the place itself is a project. The schools, homes, musical conservatories, and athletic training and playing fields can be built in ways that make the goals of the projects more or less easy to attain. The quality of the place can even extend the level of the project. With better schools we can learn more than we thought possible. The acoustics of our music halls can improve our performances and push us to compose even better pieces. The same is true with athletics. New fields and training facilities have allowed athletes to reach new heights. As place-makers we daily experience a sense of progress, feelings of accomplishment and excellence, and an awareness that what we do is subject to public inspection.

Clearly, doing a task well and seeing it done well by others involves qualities that, if not directly moral, have the potential for sensitizing us to what it is like to be moral. (Again, I want to caution that there is no guarantee.) One of these qualities stems from our realization that to attain a goal we must discipline ourselves. We also must submit to a process. We may have to learn to cooperate with others and help them in order to achieve the goals of the project. We learn to take pride in our work, and yet recognize that others too deserve credit. And we can recognize that others may have more skill, which can make us humble and yet spur us on. We can lose ourselves in our work and so provide ourselves a taste of what it is like to be a part of something greater than ourselves, that could even outlast us. If the project is done well, we feel we have provided a gift that others can use and enjoy. None of these qualities in themselves is sufficient to break out of instrumental judgments, but together they can stir us enough so that we might think that another, more general means of judging could exist.

Aesthetics as a Bridge to the Good

Increasing this possibility is the fact that these qualities move us closer to another set that is often argued to be a preparation for an awareness of the good—and that is the aesthetic. Accomplishing things skillfully and excellently usually means that we execute projects with a certain grace and economy of effort. These are aesthetic qualities, and projects that are done excellently have an aesthetic appeal. They are done gracefully and beautifully. Acquiring skills to undertake projects well can then increase awareness of the aesthetic qualities in our own efforts and those of others.

At a concrete level, every place has an appearance that can be evaluated aesthetically. At a more general level (as mentioned in Chapter 2) each place weaves together not only the domains of the moral and empirical, but also the aesthetic. The aesthetic was not included in Fig. 1 (and subsequent drawings) because the principle issue has been the relationships between the empirical and the moral. Having addressed many of them, it is now time to consider the ways in which the aesthetic domain is related to the moral. Overall, the structure and dynamics of place alert us to the fact that the elements of the aesthetic correspond to those of the moral: aesthetic judgments use ideas such as the true, the just, and the natural, which place weaves together.

As background to these complex connections, we should begin with the point that the good is always beautiful, but by no means is all beauty good. Still the aesthetic has long been recognized as something that can sensitize us to the good. Aesthetic and moral judgments have things in common. One of these is that claims about beauty and the good are stated in objective terms. Kant points out that to say that something is beautiful not only focuses on the particular object and its qualities, but at the same instant also claims that the object is partaking of a universal quality: beauty. If the judgment were thought to be subjective, we would say, according to Kant, that the object is pleasing; but to call it beautiful is to make a claim that the beauty is real. Beauty also shares with the good the sense that it is an end and not a means to an end. It pleases apart from any purpose. (But unlike the good, beauty on this count can deceive us and disguise the fact that it is being used as a means to other ends that may not be good. The good however does not deceive: it is not used for any other purpose than itself.) And, like the good, beauty pleases immediately. These similarities suggest that the aesthetic or the beautiful possesses qualities that can lead to intimations of the good overall. The connection can be more specific if we use the structural qualities of place and the idea of intrinsic judgments as a guide. Here, I will draw on Elaine Scarry's monograph *On Beauty and Being Just* to fill in some of the details.

The aesthetic presents itself not only as something complete and real, but as a gift. "Beauty," according Scarry "is a greeting. . . . It lifts away from the neutral background as though coming forward to welcome you."[87] Beauty or the aesthetic, according to Tuan, is the gift of the senses coming to life.[88] One can view it, sense it, appreciate it, learn from it, find it inspiring, but not have to possess it. That is, it can and should be public. It is diminished when owned and secreted away. When beauty is found in nature, which is so often the case, then nature too is part of this gift-giving quality that makes it seem a public trust that we all want to share.

Beauty, a gift to be shared, inspires us to create a varied and plenitudinous world. "Beauty brings copies of itself into being. It makes us draw it, take photographs of it, or describe it to other people. Sometimes it gives rise to exact replication and other times to resemblances, and still other times to things whose connection to the original site of inspiration is unrecognizable."[89] "It also inspires the idea of terrestrial plenitude and distribution, the will to make 'more and more.'"[90] This will to bring about variety, complexity, and plenitude is a gift to everyone or anyone. When a great work of art is stolen, everyone feels the loss. This is part of how beauty becomes a propaedeutic for altruism and gift-giving.

Another connection of beauty to altruism and the moral lies in the idea of justice. Beauty, which often emphasizes symmetry and balance, can sensitize us to qualities of the just. Even the words we use to describe the two indicate their similarities. Something beautiful is also called fair, which connotes treating things with care, attention, and in the right way, as we would treat people when we are fair or just to them.[91] There is also a similarity in the way asymmetry operates in the two. The beautiful often deviates from symmetry, but it seems that ultimately there is an underlying sense in which these deviations balance out; so too with the various applications of justice. One individual may require more attention than another, but this too must balance out in terms of fairness. Fairness and balance all suggest an underlying equality of care and attention that pervades both beauty and justice, and are qualities that must be open, accessible, and given away.

But is there not a difficulty here, for do we not think some things are more beautiful than others, and so when we focus on aesthetics, are we not then violating principles of equality and stressing elitism? Are not these more beautiful things more meritorious and do not the less beautiful become less important and treated with less care and attention? The answer to this is no, and the reason why is almost identical to our previous discussion of how intrinsic judgments do not lead to an elitist sense of justice. For, in the case of the beautiful, while there may be a hierarchy in beauty, all instances are important, just as all instances of the good are important, even though some are extremely good. No instance, not even the smallest, is dispensable. Nothing is unimportant in the good and the beautiful. A beautiful act helps us see even if it does not endure, for even then it opens our eyes to other instances. In Scarry's words: "it's as though beautiful things have been placed here and there in the world to serve as small wake-up calls to perception, spurring lapsed alertness back to its most acute level."[92] And even when a once beautiful thing now appears less so or ordinary, this does not mean that it is not important. "A vase may catch your attention, you turn your head to look at it, you look at it still more carefully, and suddenly its beauty is gone. Was the beauty of the object false, or was the beauty real but brief? The three-second call to beauty can have produced the small flex of the mind, the constant moistening that other objects . . . will more enduringly require."[93] In making us more aware these less powerful and more transient qualities can inspire and spur us on to greater care and attention. This is not only something that helps us, but is something we want to share and multiply, for the aesthetic sensibility strives to create a plenitude of gifts that are open and public.

Beauty's connection to justice and its spur to create variety and complexity complements its ultimate connection to truth, and the real and good. Plato and Kant saw this when they drew attention to how beauty's focus on the particular brings the general and universal to the fore, and how this leads to the experiences of certitude and conviction (and also of error) that are necessary for understanding truth. The tiniest detail of beauty, as Scarry puts it, "prompts a search for precedent which in turn prompts a search for a still earlier precedent, and the mind keeps tripping backward until it at last reaches something that has no precedent. [This is why beauty is perceived to be] bound up with truth. . . . The beautiful . . . acquaints us with the mental event of conviction . . ."[94] It inspires us to find enduring conviction, which is part of believing in a truth.[95] This why beauty and truth are allied. "[Beauty] ignites the desire for truth by giving us . . . the experience of conviction and the experience, as well, of error." For we do make mistakes about beauty, we change our minds, have debates about it, and also create and see it in infinite varieties. So when we do believe it is there, we feel a conviction, but we also can be equally convinced it is not. "This liability to error, contestation, and plurality" Scarry argues, ". . . has sometimes been cited as evidence of its falsehood and distance from "truth," when it is the case that our very aspiration for truth is its legacy. It creates, without itself fulfilling, the aspiration for enduring certitude. It comes to us, with no work of our own; then leaves us to undergo a giant labor."[96]

Beauty contains qualities of the true and the just. What about the natural? The virtue of the natural in the moral realm is also reflected in the aesthetic. The beautiful, with its grace and elegance, cannot at the same time appear to be anything but a given, a part of reality, in the same way that the givenness of nature makes the natural into a virtue. The true, the just, and the natural then are all present in aesthetic judgments, and when the aesthetic and the moral are literally placed together and complementary, the aesthetic (like the moral) becomes expansive. It draws attention to the vastness and complexity of the world. It makes us aware of the general while still attending to the particular. It is a life-enhancing force. In *Passing Strange and Wonderful*, Tuan says that "at a general level, good is simply a fusion of moral and aesthetic conceptions, which include a sense of rightness and appropriateness, care and accomplishment, delight in the way things are done and in the things (both natural and artificial) themselves."[97] Such sensibilities are cultivated by our acquisition of skills, and in projects well and beautifully done.

The moral and the aesthetic, with their elements of the true, the just, and the natural, and the connection of beauty to altruism, all make sense in the context of place, for it is place-making that weaves these together. This is true in the abstract and in the concrete. The weave can be found even in the most ordinary place. The classroom, for example, not only contains elements of truth, justice, and the natural, but also an appearance that can be aesthetically evaluated in corresponding terms. And if a classroom that was well-appointed were also the site of lessons on the dangers of hate, the aesthetic environment would enhance the message; indeed, the message itself would make the place appealing. But if we do not keep the real and the good in mind in this kind of place and in all others, both duty to a project and doing it well and beautifully can all too often become subverted, and serve as a veneer to disguise immoral practices. A Nazi classroom may be beautifully appointed, but this aesthetic quality cannot for long disguise the ugliness of lessons of hate taught there.

All too often, the aesthetic and the moral are not working in consort, and beauty is a product of lies and injustices. So many grand aesthetic projects like Versailles, St. Petersburg, and the construction of innumerable palaces, gardens, and temples elsewhere around the world, were accomplished by displacing former inhabitants, by pushing slave and corvée labor beyond endurance, and by a callous disregard for natural habitats. The radiance of these places is dimmed by the darker side of their histories. Nazi Germany, more than any 20th century political system, attempted to fuse the political and the aesthetic. It was driven in part by an "aesthetic sensibility." The German *volk* was supposed to embody beauty. The pure German landscape, expressing this beautiful spirit, was to be itself a work of art, with pleasing and comely vistas, and tidy and orderly German farms, fields, and villages. The entire country was to be unified around monumental works of art that were stage settings for the grand mass rallies.

The aesthetic within the context of particular places and projects alone then is not enough to assure that we move in the right direction. Indeed, the aesthetic can lead in the entirely wrong direction. But the point here is that it does have qualities that sensitize us to the good. As an ideal that transcends any particular project, the aesthetic shares some of the qualities found in intrinsic geographic judgments: a dedication to opening or eyes to the real and valuing the complexities of the world.

While not all beauty and excellence is good, anything good is always excellent and beautiful. As a guide to the good, intrinsic judgments can help make the instrumental more like the intrinsic, and thus make the excellence and beauty of instrumental activities good. Intrinsic judgments remind us that the excellence and beauty of good places and their projects are public contributions or gifts. They become part of a more open place-making and a countermeasure to the narrowness of moral drift and self-deception.

The boundedness of place—its compartmentalization—is a pivotal part of the geopsychological dynamics. To use it correctly depends ultimately on our possessing a strength of will to be open and aware of the good. Our emphasis on the moral does not mean that the empirical and especially the social are not important. We have all too often seen how difficult it is in many social systems to see clearly and create variety and complexity. The significance of social relations is in how they help or hinder these capacities. The next chapter will consider the kinds of social institutions that would help. Rather than following a process of small steps where improvements to place lead to improvements in social relations, that in turn lead to improvements in place, that improve society, and so on, Chapter 7—Geosocial Dynamics—jumps ahead and presupposes that we already accept intrinsic judgements as a guide, and then considers the kinds of social relations that are animated by and support these judgments. The altruistic implications of intrinsic judgments—that truth, justice, and the natural are gifts and that good places impart them (thus making these places beautiful and excellent)—will be the focus of attention in the next chapter. I will make the case that altruism and altruistic gift-giving are at the core of a morally justified social system and the primary force that circulates within its political and economic networks and institutions.

7

GEOSOCIAL DYNAMICS

OUR MORAL SENSIBILITIES URGE US TO USE our free will and become less situated and more aware. Yet it is easier not to be. And even if we want to, it is especially difficult if the rest of society uses place to escape and drift morally into a narrower set of concerns, so that its members become self-deceived. Then, any particular individual's efforts to do otherwise would be heroic, and, what good effect might come from it, all too limited. This is why we all need the assistance of good social institutions and social relations. But even then good institutions do not produce good without us having a will to do it.

A focus on free will allows us to see that social institutions, social structures, and the forces they exert, even though real and powerful, are the (often unintended) consequences of our own decisions. They still affect us, but they can be changed. They are not laws of human nature, but regularities that result from complex rules we follow. These regularities and the rules that generate them can be accepted or changed into what we think ought to be. With this in mind, I want to sketch the connections between intrinsic judgments and social structures and relations. I will do so by focusing on the type of social relations—or more exactly, the social, political, and economic relations (all of which are subsets of the realm of social relations)—that intrinsic geographic judgments call upon.

The social, the political, and the economic are of course intertwined, and place draws interconnections that do in fact cut across and alter the dominant paradigms of each. I will take these in order, beginning with the social as a general category, and then the political, where I will focus on democratic theory, and then the effect that a stress on altruism and altruistic gift-giving would have on the economic under both capitalism and socialism. Again, each of these is hardly a discrete category, and many of the points will blend through.

THE SOCIAL

The theory does not tell us that we must have houses, or factories, or schools, or nation-states. Rather, it says that we must have places, and whatever they are, they will have this loomlike quality and should be used according to the criteria of intrinsic judgments. To offer a list of what types of places would or could be good, is to close off our creative potential as place-makers, and to run counter to the intent of intrinsic judgments. Even with this openness in possibilities, there still will be some underlying relationships (among the ones discussed in Chapter 2) that can be expected simply because of the structure and dynamics of place. The one I will focus on here is the problem of scale, for it has much to say in the subsequent discussions on democracy and the economy.

Scale

It will be recalled that while scale relationships are contingent and derivative of the structure and dynamics of place (and especially its rules of in and out), there is one type that may appear to be necessary because it is so simple and found so often. This relationship stems from the fact that the existence of several places near each other and that may have absolutely nothing in common but proximity will lead to the need for a new level of place to coordinate the unintended spillovers and other spatial interactions that result simply from their proximity.

For example, a person may be able to work at home through the Internet. The home becomes a place devoted to this and other projects. Next door is another person who also is working at home through the Internet. The two individuals, their households, and the projects they are undertaking may have absolutely nothing to do with each other; they may never even have met. Still they are connected by the fact that they are near each other and so they require and use the same infrastructure, and each creates spillovers that affect the others. The individuals in these two places have common and overlapping needs in the sense that they require roads, sewage systems, power grids, police and fire protection, schools, shops, and so on, and these activities in turn must be supplied and coordinated by a second type of level or scale of place—a political or administrative one. So, the need for the administrative unit arises simply because of the proximate relationships among these particular places. But once it is created, it then initiates new rules and effects. This second geographic level concerns the need to have one or more places that are responsible for coordinating necessary geographical flows and relations among the first kind, to provision the infrastructure, and to provide a common place to which the first type of places and their occupants belong. But again, once the second level arises, these create new rules and relationships affecting the first.

I have described the second level places as political or administrative, but I want to refrain from calling them that now, because these terms may have us overlook other important qualities of the relationship. In order to spell these out, I will refer to both types of place more generally as the first and second type or level.

Almost any place can contain other places. A factory and office building may contain particular subunits and a home may have rooms. This fact alone does not make them two types of place in this sense unless there are rules stipulating their interconnections. Consider the home. It can become an example of the second type of place, and the rooms, examples of the first, when we think that there may be "house" rules that apply to its rooms. This of course is not the way we usually think of a house and its rooms and flows among them—for both levels are so intimately involved in the same projects—but my point is that we could do so, and in fact are pressed to do so when there are others who have different ideas about what should take place in a house. A child for example may decide to cook in a bedroom rather than in the kitchen, in which case a parent may invoke something like a "house" rule, by saying for example that "this kind of thing is not done in this house."

The two scales become more distinct at the level of a village, and even more so in a town, and then in a city where there are multiple levels of the second type, and so on. In a village for example we find that there is a bounded area—the village—that not only contains houses, inhabitants, and streets, but also is an area that is somehow actively responsible for some of the relationships among them that result from their geographic proximity. That is, the houses and paths, or streets (if it is a modern village), are part of and in some sense "belong" to the village, as do the inhabitants of these houses; and some or all of the members then contend with these communal relationships. The place then becomes an instrument to interconnect and coordinate people and other places within, and if some of these were to change, the integrating and coordinating place would still be there. Clearly, as I noted, this second level has a strong political ring to it, but I want to recognize that it also refers to a more general and social geographical idea: that individuals and their projects are part of *territorially based communities*. This second level is the place-based or territorial community, and there will always be these—this second level of places (or provisional wholes)—to which those individuals, social units, and places within belong and through which their proximate and shared geographical activities are coordinated, though not always voluntarily and happily. That is, the second level will always serve in some degree as a means of identifying and classifying membership in a community.

The activities that the second level addresses can never include all of the activities and interconnections generated by places and their projects. I do not mean simply that activities will unintentionally spill over the boundaries of the second level and thus would need to be contained and coordinated by an even larger place of this kind—a state instead of a village. Rather I mean that there are and always will be intentional spatial flows that are targeted to other places outside and far away. These flows can form nonterritorial communities—ones defined by other criteria than by being part of a place of the second type. In my own case, I am a teacher, and so I have something in common with others in this group or "community" wherever they may be; and this relationship can be made more formal, as when I become a member of an international community organized to promote the study of geography. All societies always contain a mix of both types of communities. It may be the case that a greater percentage of membership in recent times is of the nonterritorial sort, but even on this account I am not sure, for there are now so many of both. Not only am I a member of, or identified with, many nonterritorial communities, I may also be part of multiple layers of territorial ones—from the neighborhood to the city, state, and nation. Moreover, I find that the nonterritorial type often becomes territorialized. For example, although my association as a teacher of geography does connect me to many other geographers around the globe, the association defines members, and organizes their connections, territorially—I am a member of a regional association, an American association, and international one. All cultures then will have both types of membership, and there will always be these two levels of place.

Places of the second level are not simply political instruments, for they involve other social functions and identifications (and they contain important moral implications which will be addressed shortly). But political administrative issues are often among their functions, and this leads us to examine the way they are embedded in concepts of political community and principles of governance.

THE POLITICAL

The distinction between the two scales suggests that the second level is most directly linked to what is usually described as the political. This is partly true, and to understand why, we must step back and think about the general relationship between place and power. Many would argue that the very act of creating and sustaining any place is political, if the political is about how power is used and justified. Deciding and enforcing what should or should not take place presuppose positions of authority, decision-making processes, and power. These can be informal or formal, customary or novel, accepted and taken-for-granted, or highly contested. Whether they are called political or not, intrinsic judgments lead to the idea that all forms of social power must somehow derive from democratic sources and processes. These democratic processes should then extend beyond the conventionally political. Intrinsic judgments also make a distinction between an instrumental and an intrinsic use of democracy, and most definitely encourage the latter, for that is a democracy rooted in altruism and altruistic gift-giving. Let us turn to these two views of the nature of democracy, for they draw upon the most important political implications of intrinsic judgments.[1]

Instrumental Democracy

Qualities of democracy change as it moves along the instrumental—intrinsic continuum. We can be at the extreme end of the instrumental and adopt something like an emotivist or egoist position. In this case, a person wants a system of government to satisfy his or her needs. I may think that being a dictator would assure this, but it is a risky all-or-nothing strategy for me. A safer bet is to settle for some sort of balance of power where my interests will have the same weight as everyone else's. A democracy, with checks against the tyranny of the majority, in theory can provide this. "One person, one vote" then means that my interests have as much right to representation as do yours, and "interests" mean situated, self-interests. Along these lines, in a democracy based on elected representatives, these individuals then represent the interests of their constituents, and I would be wise to vote for a representative who will best further my interests. The democratic structure then becomes an arena where interests are voiced, conflicts arise, and agreements reached through alliances, negotiations, and comprises.

A deeper justification for this view is found at the heart of what might be called liberal democratic theory. This sees democracy as a means of arriving at a fair and just society that promotes individual rights. The basic idea, according to Michael Sandel, lies in the assumption that "a just society seeks not to promote any particular ends, but enables its citizens to pursue their own ends, consistent with a similar liberty for all"[2] and that the best system to achieve this is democracy. Democracy allows each of us to define and pursue our own interests, and the state will be neutral about interests in general. It will not pick sides or promote one set over another. The state leaves the question of "What is good?" for each of us to decide, and does so by guaranteeing our rights to liberty—to pursue our own views on this matter and to pursue our own interests. These rights to liberty are equivalent to assumptions about our rights as autonomous agents, and the need for all of us to respect these rights in everyone else. What is being guaranteed then is a process stemming from rights, and this is where justification stops. It does not go farther to justify why these rights are really good, or to ask how anything good can come of them. It claims instead that rights precede the good. We cannot discuss the good without first having stipulated rights.[3]

The difficulty with these instrumental views of democracy is their incompleteness. How can we know that these rights are *right* without some inkling of what is good? If we have this inkling, it behooves us to make it clearer. We need a conception of the good to know what rights are *right or good*. Without it, our supposed duty to adhere to particular rights rests on a shaky foundation. Let us turn now to how intrinsic judgments can be linked to democratic theory to forge an intrinsic concept of democracy that builds support for and extends these issues of freedom, equality, and rights.

Intrinsic Democracy

The theory builds on and extends the instrumental view so that democracy does more than guarantee and promote rights and interests. The main argument I will make here is that in the hands of intrinsic judgments democracy becomes a socioeconomic mechanism for promoting altruism and for moving toward the good. The argument entails two smaller ones: it is unlikely that democracy can be used in this way unless we decide to (and that requires that we assume the good exists and construct places that follow intrinsic judgments); and that once we so decide, the conventional liberal conceptions of rights associated with democracy—namely, the human rights of equality and the freedom of individuals to decide what is good—are not only preserved, but supported and given a purpose. That is, the same rules and rights are maintained but given moral vitality. The political system then becomes an arena for giving and sharing with others the attempts at seeing clearly and of creating variety and complexity. I will begin with the second entailment: that the liberal view of rights is enhanced and given purpose.

A liberal theory, based on the autonomy of the individual, supports rights and condemns individuals who decide that the rights of others should not be recognized. Without assuming the existence of the good, the liberal model's emphasis on rights appears to be unsupported except as a means of assuring self-interest, and this leads to difficulties. One is that preoccupation with having my rights guaranteed could in fact make me care less about being an autonomous agent. As long as the rights are provided, I do not question where they come from, or my role in assuring them. I may want the right to life (including food and shelter), security, a (good) job, the right to medical attention, and so on. If these rights can be guaranteed by some beneficent dictatorship, that may be enough. A dictatorship that is running smoothly, eliminates poverty, provides stability, and guarantees other rights can come to have the support of its citizens. If all goes well, the dictatorship can even allow the rights of free association and speech, and even some degree of political participation, for it has little to fear from its citizens. This of course is most unlikely, but an emphasis on rights and a narrowing of reasons to have them can encourage individuals to become passive recipients, and not essential parts of the political community.

Related to this problem is that, even when people are actively participating, an emphasis on rights and self-interest may make it more difficult to think of obligations and duties. Why should I ever vote for something that is not in my interests? Why should I ever want to raise my taxes if I personally am not the beneficiary? Why should I want to encourage public education if I do not have school-aged children? Liberals have of course argued that such concepts of obligations and duties, or positive rights (as they are sometimes called), as opposed to only negative ones, can follow from an enlightened self-interest. They can be shown to eventually accrue to the benefit of the giver, by creating a more stable and appealing society in which she or he lives. But then if the benefits do not flow back, the giving stops, or if the benefits can be had without giving, then there is no need to give. Another risk to a focus on rights is that when we only think of receiving rights and not giving or making contributions, we may come to think of ourselves primarily as dependents, and this diminishes our own dignity and self-worth.

Connecting liberal democracy to the kind of altruistic impulse embodied in the theory helps diminish these risks. But it does more. The free and open aspects of discussion which are part of democracy and intrinsic judgments become not only the means of giving voice to one's interests, but a means of contributing and sharing so that the collective likelihood of being aware of the real and the good is increased. Democracy becomes a form of public conversation and exploration of different views that does not have to be confrontational, but rather mutual, and even generous. Democracy becomes a means of arriving at a less partial and more enlightened position. This connection between democracy and enlightenment is part of the position of Jürgen Habermas and John Dewey.

Habermas argues that the world can become more just through sifting and winnowing our views in a free and open discussion or conversation (which he calls an ideal *speech act* that takes place in an ideal *speech community*, to differentiate it from other kinds of speech acts such as commands).[4] He finds a deep connection between these advantages of open conversation and the role of democracy. True discussion presupposes an ideal speech community that is virtually equivalent to an ideal democracy. This speech community is one in which all individuals have an equal chance of initiating and responding to and questioning assertions in a coercion-free environment.[5] This democratically based system of communication encourages a view that is more detached and less self-interested, because the democratic quality encourages us to put ourselves in the position of others. Habermas views democracy as "a self-controlled learning process."[6]

Democracy then is a means of expanding our awareness and having us become less situated and is integral to a moral way of life. John Dewey's philosophy of democracy, though based on pragmatism (and offered more than fifty years before Habermas), also urges us to be sensitive to the interests of others, and to see democracy, education, and geography as an interrelated way of providing a less self-centered view. Dewey believes that democracy is a conversation that seeks to enlighten and educate, and that these associations lead to a shared sense of humanity.[7] This commonality will embrace everyone, without geographic limit.[8] Dewey makes the point that education, democracy, and an expansive morality, not only converge, but that geography's place is essential to this process because it integrates the sciences and expands our horizons: "it is an intellectual starting point for moving out into the unknown, not an end in itself. . . . To follow [this] course is to enlarge the mind . . . by remaking the meaning of what was previously a matter of course."[9]

These are important points where Dewey, Habermas, and the theory converge. But Dewey and Habermas hold conceptions of truth that differ from that of the correspondence view assumed by the theory. In Dewey's case it is a truth based on what works, or a pragmatic theory. In a different way, Habermas also tries to avoid a commitment to a correspondence theory (and by implication may even hold a consensus theory, though it is not made explicit) in that he sees democracy and conversation as not only necessary but may even be sufficient to resolve questions of truth. He suggests that whatever everyone agrees to under the conditions of a democratic and ideal speech act is all there is to truth. While a correspondence theory would support Habermas' view that democracy and conversation are extremely important and even necessary for comparing, discussing, sifting, and winnowing, which are processes in the discovery of truth, it still holds that conversation and democracy are not sufficient. We also have to have a reality that draws our attention and that helps us test, confirm, and reject even those ideas that are wrong in spite of the fact that we all hold them or, in Dewey's case, that we find that they work but do not know why. These differences aside, the intimate connections among intrinsic judgments, democracy, education, conversation, and the search for truth remain.

An all-important point is that these pursuits are, and take place in, public. What then does the theory say about the public and the private? How does the requirement to create places and projects that heighten our awareness and create a varied and complex world—all of which is then given publicly and buttressed by an intrinsic democracy—justify anything but the public, and what bearing on the matter have the two levels of places we discussed in the first part of this chapter? The best way to address these issues is to compare intrinsic judgements with yet another view of democracy—that of radical democracy and its conception of public place.

Radical Democracy, Intrinsic Democracy, and Public Place

On the surface, intrinsic judgments and radical democracy have much in common. They join in recognizing the dangers associated with power and the difficulties of power relations in a democracy. The two share the view that democracy must penetrate than realms of life other only the narrowly political, and that it is a principal means of critiquing power. Yet they differ fundamentally over the role of free will and self-interest. Radical democracy does not explicitly recognize the capacity of the individual to become less situated and more altruistic. Because of this it is encumbered with the same problems of postmodern deconstruction that we discussed in Chapter 3. Specifically, it does not provide a guide to what is a good use of power, to the possible limits of democratizing everything, and so, by *default*, presents a chaotic political outcome of groups constantly vying for power, and suggests that this continuing and destabilizing quality will eventually be emancipatory. Its geographic implications then are similar to those we described in our discussion of postmodernism, and these crystalize around the role of public place.[10]

Public place Public place, both actual and metaphorical, plays an important role in radical democracy's critique of power. Yet its conception of public place (and the role of place in general) is clouded by radical democracy's inability to take a stand on the existence of an autonomous self. Intrinsic democracy makes it clear that a public place is a gift. It requires those participating to voluntarily relinquish degrees of self-interest. But radical democracy's focus on situatedness cannot lead us in this direction. In spite of this, many advocates of radical democracy yearn to have true public places where conventional power relations become swept away so that something like a purer emancipatory power, or at least its potentiality, will fill the void. The controversial point here is not about whether we need something approaching true public place. Rather it is that while the various models of radical democracy assume on the one hand that such a place is important to establish, they don't explain why, and many seem to despair that it can be created because everyone's position is situated. They fail to recognize that place itself, including public place, requires the existence of some stable set of rules based on the application of power, and that the only way to do this so that a public place results which is open to everyone and in which discussion, debate, and critique of power can take place (which are the goals of both radical and intrinsic democracy), is for those involved to agree to reduce the role of self-interest—which in turn assumes the possibility of free will. The existence of public place itself becomes a continuous entailment of altruism. Because they are not addressing these issues, radical democracy's discussion of public place tends to be muddled.

For example, Rosalyn Deutsche sees only self-interest as a motive, and yet wants those motivated by it to debate their interests in public. How then we can reduce our self-interests to form a public place is left unexplained.[11] In a more extreme and perhaps consistent version of radical democracy, Nancy Fraser and Bonnie Honig refuse to admit that "a" public place can emerge.[12] Rather they argue there can be different public "places," each coinciding with the self-interests of its members (which makes them hardly public). Others though, like Claude Lefort, are less committed to an entirely situated self and more optimistic that a single type of public place can emerge. They believe democracy can work to join diverse groups if it is thought of as a continuous and open-ended process enabling all of us to challenge and debate about rights. This seems closer to intrinsic democracy. But it still does not examine sufficiently why the right to question is good, how we move from a rights-based discussion to a less self-interested form, and why this can and must occur in a public place.[13]

These positions assume that public place (or places) is central to the project of radical democracy, and more generally to democracy itself. They do not however explain either how it can exist when most everything is subject to challenge or how it has an effect. It seems as though all of the important but unstated assumptions about the good are being projected onto, or conflated with, this unexamined site. To unpack how public place works politically, we need to build on the general qualities of the structure and dynamics of place, and also connect these to the intrinsic assumption about the possibility of places promoting a less situated and more altruistic position.

This is the direction Hannah Arendt's political philosophy would have us go. She is able to fill in some of the details about the need for, and function of, public place in political life because she sees the political as more than a confrontational meeting place.[14] She assumes the public to be those literally gathered together in place to undertake cooperative actions (even if this action is to debate). Though members of the public may have entered with different points of view, their collective political decisions now allow them to share an enlarged, less situated, and more public awareness, as well as a common set of actions and projects; and these, like any other undertakings, require place. Place and the public, as a political body, are mutually constitutive, and one of the essential acts of the public is to initiate and maintain this place. As she says: "The space of appearance is created every time individuals gather together politically, which is to say, 'wherever men are together in the manner of speech and action,' and in this respect it 'predates and precedes all formal constitution of the public realm and the various forms of government'. . . . It is a creation of common action and collective deliberation, the space of appearance is highly fragile and exists only when actualized through the performance of deeds and the sharing of words."[15] It is the world we create and hold in common. This public arena enables the political because "political opinions . . . can never be formed in private; rather, they are formed, tested, and enlarged only within a public context of argumentation and debate."[16] Participating in public place making, as well as sharing in political values, are the ingredients that unite people.[17] Through the public we transcend our more limited and private worlds.

These ideas are perfectly consonant with those of intrinsic democracy. They point out that just as politics is a process, so too is place construction; public place helps construct, and is further elaborated by, the activities individuals engage in collectively—not simply confrontationally.[18] These activities include undertaking physical, social, and intellectual projects. And if people come together democratically in the sense of being open-minded and willing to contribute, and if they focus primarily on the realm of meaning, then they can constitute, in Habermas' terms, a speech community, that allows its members to grow intellectually by sharing ideas. These public places enable gift-giving and they themselves are a gift—enabling a variety and complexity of ideas, and the capacity for us to share them and see reality more clearly. This is why the connection between public place and democracy can be better understood when examined in light of intrinsic democracy.

We can be more precise about the type of public places and the way these are related to the private. Let us consider again the two levels of place that were addressed in the first section of this chapter. As we noted, the two would exist in any society, whether democratic or not, and the second level would be most directly related to the political in a conventional sense even when it is only an administrative area (which under intrinsic judgments would mean it must be thoroughly democratic). Intrinsic judgments would argue that, guided by "seeing through to the real" and "variety and complexity," the second-level places will be thoroughly public in that the workings of place—the operation of the loom, knowledge about the elements woven, the rules of in/out, issues of appearance and reality, the flows, and who is doing the weaving—should be made as transparent and visible as possible, that the place would be thought of as providing a contribution through both its function and in its very landscape, and that this contribution be accessible to those outside as well as those within its borders. These criteria for the second level would meet the ideals of public place espoused by some radical democracy advocates.

What of the first level? Because of its emphasis on gift-giving, the intrinsic democratic theory encourages the first level to become closer to the second in its democratic and public qualities in two related ways. One follows from the fact that internalizing intrinsic judgments can increase the likelihood that places at the first level encourage the quality of gift-giving and contributing, which will make them less instrumental, and us less situated. Even in a capitalist system privately owned places will be thought more of as places held in public trust with a public duty that does not necessarily end at the border of the second level. Again, this applies both to the products of the places and to the places themselves. That is, their landscapes become part of the ensemble of public contributions.

The second sense follows from the fact that when intrinsic judgments are internalized by all of those who are making places and projects, these individuals must also be involved in assessing what is taking place. This can happen only if the structure and dynamics of place are open, public, and accessible. In many cases, this may mean that these places should be thoroughly democratized. I say in many, but not in all. This is because intrinsic judgments and intrinsic democracy recognize that we not only should question, contest, and keep decisions open, but also that this must be balanced by the need to allow places meet the criteria of intrinsic judgments to persist so that they can fulfill their potential and we must also recognize that some of these places cannot function if their decision-making rules are thoroughly democratic. It may not be possible or good to run a family completely democratically, nor perhaps should a school, or a hospital, or many other places that require specialized knowledge and particular social obligations and responsibilities, be thoroughly democratized.

This does not mean that the voices of those engaged in the projects will be silenced. Rather it means that if the place is sanctioned by a thoroughly democratized second level, then it also must be given the right to be a somewhat separate sphere with its own mixes of virtues. Otherwise, we may become engaged only in contestation and deconstruction, and we also may well prevent the possibility of having places that are varied and complex. I will take up this Walzer-like issue of separate spheres again soon, but as a final point, we must remind ourselves that no system will work unless its values become accepted and internalized by its members. So, again, I am assuming that people have internalized intrinsic judgments and use them to guide their own actions. Once we decide to take up these principles, the political procedures and the places become inflected by these values and help support them.

Internalizing intrinsic geographic judgments makes the public quality of decision-making at the first level more like the second. But there is a subset of places at the second level that is intensely public and deserves special attention. I am referring here to places that are set aside for the public to enter and gather solely as members of the public. The purest example of these are places where people meet to do the work of the public—the parliaments, courthouses, and other places of assembly and government. Another is a place that represents public works: works that the members of the community commission or construct themselves, and which can also become places of public assembly. This can be the case with streets, town squares, parks, and nature preserves. Even when these are not being used directly for assembly, they can still provide an intense idea of a public place by offering an opportunity for strangers to simply share a place that is in part theirs—to offer one another public hospitality.[19] They also are public creations, and so their products and appearances are gifts from the community. These ideas can and should motivate the first type of place, even when they are owned and operated privately, for they too are linked to the same landscape.

The conjunction of democratic theory and intrinsic geographic judgments would argue for an expansion of such places and their uses. These are the places that are most important for the public activities described by Arendt, Habermas, and Dewey. These must be real—not virtual—places of assembly, contact, and hospitality, because to enter a real one requires traveling through, sensing, and observing other real places and activities (at both geographical levels) and being in the presence of strangers whom we cannot screen out, and with whom we must share a place. The experience gained by these encounters with landscapes, and the sharing of place with strangers, is something that cannot be electronically simulated or captured accurately by symbolic systems. These are too selective and reduce the experience to only a single realm which, while being in real place and moving from one to another, not only forces us to encounter the structure and dynamics of place, but also engages us in their maintenance and their change. They help create, what some liberal theorists have called, a civic culture. Yes, they provide a place for us to be antagonistic and agonistic, but from the intrinsic point of view, they are public creations that impart the gift of enabling us to meet, share, care for, and learn from one another.

The levels of place and the role of the public are one important arena illustrating the geographic assumptions of intrinsic democracy. There are many more. Here I will briefly mention three—areal representation, membership, and non-democratic spheres—that explore further issues into these two levels.

Areal representation   Democracy is designed to allow individuals to govern themselves, and each individual is allowed a voice and a vote. When the system is complex, individuals require representatives. What does intrinsic democracy say about representation? The point that will be made is this: representation should never violate the principle of one-person, one-vote, but the system of representation must be designed so that some, if not all, of the representatives are elected by and represent members of an area, district, or territory. This is often called *areal* or *territorial* representation and is contrasted to *at-large* representation where voters anywhere vote for the same slate of candidates. These candidates then do not represent a geographical group, but the entire population. The necessity for areal representation follows from our discussions of the inescapableness of the second level and its role in defining activities and communities. When electing a representative by area, that representative is representing the constituents as individuals and their joint geographic connections and landscape. At the same time that a school or police or fire district, for example, coordinates specific functions, it also imposes on its residents the properties of belonging to a community. It creates, as we said, a territorial definition of social relations—a territorial identity: individuals, solely by virtue of being within a boundary, share certain interests, rights, and responsibilities that will be expressed in the landscape. These communal relationships may be as thin as being part of a school district, or they may be far thicker, as when a city or state molds activities and identities. Whether small or large, thick or thin, these units are a part of life and have important consequences. Their boundaries and weaves will be contested and altered, but their presence is inescapable.

An individual then is part of these territorial communities. But he or she is also part of nonterritorial ones. Indeed, identity politics reminds us of how many ways both types can be used to define and redefine ourselves. How should this mix of territorial and nonterritorial concerns be reflected in the structure of political representation and the slate of candidates for whom one can vote? It is possible to represent only by area or only at large—where voters from any place can vote for any candidate (the any place though is still usually within some larger territorial unit, like a nation-state, that defines the polity as a whole). As I said, the theory argues that areal representation is essential, though it does not have to be the exclusive form.

The theory cannot be more specific than that. It does not stipulate the geographical size or how many levels of these areal units there should be; but it does say that it should not violate the one-person, one-vote rule.

Membership Areas should be represented, but who does an area include or exclude? If an area affects others outside, should they not have a say in what takes place? If they do, are they not in some sense political members of that place? Membership is a central problem in democratic theory. Democratic systems are predicated not only on the assumption of one-person, one-vote, but on the assumption that those affected by political processes should have a voice or a vote in them. This is a problem, for no areal unit is self-contained. Places are interconnected and affect one another, and at an ever-increasing rate. This is true for both levels. Activities such as manufacturing involve networks with other places outside a particular political territorial unit, and administrative and political territories create activities that spill over their borders. One territory's policy about labor practices can affect the type and location of manufacturing in another; policies about ethnic minorities can affect immigration; a territory's environmental policy can affect every other territory around the globe. What then ought to be the rights of political participation of those who are affected by, or interested in, what takes place in another territory? What then is membership, and who belongs to this or that political unit?

One means of addressing this problem of membership outside a territory's borders is to extend the borders and the number of hierarchical units in the second level. The geographic spillovers, the nonterritorial forms of community, and the general push of intrinsic democracy toward altruism certainly argue for increasing areal coordination that would eventually be global. They encourage a flexible and porous concept of sovereignty that can lead to a hierarchy among territorial units to coordinate these spillovers, connections, and contributions. The theory does not however stipulate what the details of the second level of places should be apart from the fact that they must be democratic and guided by intrinsic judgments.

Another way of discussing membership is to consider whether people can enter and exit these areal units. The theory would encourage open boundaries or, in terms of current territorial states, open immigration. That is, not only are the boundaries of the political units arbitrary and changeable, the members of the political units are free to come and go and become members of others.[20]

Those who want to become members enter because they want to contribute. This means they must accept the systems in place to some degree. Let me explain. By system I mean not only a democratic process, but a cultural way of life that may involve a particular language and set of beliefs. That is, it involves all of the qualities that are part of places in the second level. To want to settle in such an area means a person finds these activities not only attractive but wants to become involved and make a contribution to them (and as we shall see in the next section to places of the first level too). In doing so, that particular system and way of life will also be changed, for those who arrive would have the right to participate. They, like any other citizen, can help preserve or change even its deepest cultural characteristics. Wanting to participate, to make a contribution, is an important requirement of being accepted.

If entry and exit at the political level is open, then issues about boundaries, responsibility, and membership become more diffused. Still, there probably should be some requirement that has a person be a member of only one political community at a time. There are several reasons for this, but perhaps the most compelling is that free entry and exit can thin out our sense of obligation and participation in an area. We may not feel part of and committed to any particular place or project. Commitment to places, projects, and ways of life is important. Places are inextricably linked to what we make and contribute to the world. We must stay long enough to commit ourselves and yet remember that we do so freely and that this commitment does not irrevocably determine who we are, or prevent us from leaving. If we are motivated by intrinsic judgments, it seems that we would also impose on ourselves the stipulations that we leave only if we do not diminish the present contributions, and if we can help create others of equal or greater value elsewhere.

Spheres   This idea of free entry and exit would apply to places of the second level. But what of places in the first? Should anyone who is in a territory also be allowed to become a part of any particular place and project within it? And, is not entry into the second usually dependent on having a place at the first level in which to work? Certainly, the issue of entry at the first does affect whether people will want to enter a territory, remain in it, and contribute to it. While intrinsic democracy cannot fine-tune these relationships, there is the overarching point that places in the first level are not as open as the second. First-level places often require particular talents and skills that fit the projects taking place, and these places also involve many nondemocratically organized projects. Engaging in places at the first level means also submitting to a certain discipline and authority that comes with an art, a science, a technique, an education, a sport, a game, or any project at all, and people do so because that is how they can make a contribution. Through democratic oversight, a democratic system at the second level grants the right of these projects to take place and with their own criteria of membership or admission. In so doing it requires that the loomlike quality of these places be as transparent as possible and that opportunities to acquire requisite skills be available to all. Still, there will be differences in what is acquired. One has to meet certain standards of training and skill to perform certain functions, and these again are not necessarily determined by democratic forces at the second level. This is why there cannot be the same degree of free entry into places of the first kind as there is to places of the second. But this is not necessarily a conflict, for intrinsic judgments would assume that a person would want to move to a new place because the person could be engaged in projects and make contributions to both levels.

THE ECONOMIC

Thus far, I have attempted to show how we might move along the continuum from a more to a less self-interested position within the context of the social and political. Now I would like to do the same for the economic. I want to understand how this sphere can be animated by a less self-interested and altruistic impulse and, more specifically, by what I will call *gift-value*. Here, the problem of drawing connections to the theory is far more difficult primarily because the principles of self-interest have a firmer and more explicit grip on economic models. Modern economic theory holds that it is rational for individuals to try to maximize their self-interest (and conversely irrational not to) and that this process can be quantified because units of money can stand as a measure of self-interest. The pursuit of self-interest, coupled with the mechanism of the marketplace, is supposed to lead, through something like "the invisible hand," to an efficient allocation of resources and the best good for all.

Clearly, intrinsic judgments see it differently. The theory would argue that holding this self-interested quality as the model of economic behavior, and leaving out, or explaining away, or belittling as irrational the essential human capacity of being less self-interested and more altruistic, would diminish our sense of self, truncate the moral to instrumental judgments, and make it necessary for those who wish to act in a less self-interested and more altruistic manner to claim their motivations to be non- (or) extra-economic. To avoid this, we should consider how to make intrinsic judgments and altruism part of an economic model. How can we do this? There are two roadblocks to overcome. One is that altruism, and its economic formulation in terms of intrinsic value, is not capable of being quantified without distorting or diminishing the concept. Gifts are unique and their values "grow" as they become open and accessible to all. So how does one connect such a mostly qualitative idea of value (as in gift-value) to a quantitative one? This leads to a second difficulty—that of coordination. If altruism becomes a strong motivation to engage in work and production, and if it is not readily quantifiable, then how can creations and contributions be coordinated? If our motivation is altruistic, how can we and others be confident that we can still make a living?

I cannot claim to have specific answers to these points—and in this respect I confront the same difficulties that a Marxian utopia faces, which is that there may well be conflict between the self-expression and actualization of an individual, and similar attempts by others in the community, society, or world. All I intend here is to contrast the idea of gift-value that comes from intrinsic judgments to other ideas of value (that are inspired more by instrumental judgments), and to suggest that perhaps a blending of these ought to be pursued. This blending has the best chance if we remind ourselves that accepting intrinsic judgments amounts to having the joint values of "seeing through" and "variety and complexity" become our most essential and critical needs. Since meeting these personal needs is accomplished by also making contributions—open, public, and accessible to all—then even though we may assign value on one or another of these offerings through something like conventional marketplace forces, these quantitative relationships of value will be influenced by the primary and qualitative ones of intrinsically judging and valuing the products, and also the processes of producing them, in terms of how they meet our needs to create open and accessible contributions and insights about the real and the good.

Before I discuss gift-value and other types of value, I need to consider if altruism (or intrinsic judgments in general) is more closely related to one or the other of the two major economic systems: a Marxist-inspired socialism/communism, or capitalism. The point I will make is that while both employ strong assumptions about self-interest, there is more room in socialism (and market socialism) for altruism. Even so, the socialism inspired by Marxist theory does not allow altruism to flourish because it does not take free agency seriously enough.

Capitalism and Socialism/Communism

Capitalism and socialism/communism are complex ideas and practices with numerous variants. (For the sake of simplicity I will use the terms socialism and communism together, unless the context warrants a distinction.) Our interest is in the degree to which either of these broadly understood economic systems could be used to encourage the values of altruism and gift-giving.

We have already noted that the theory of capitalism has come to be inextricably bound up with the concept of self-interest.[21] In its classical form, proponents of capitalism explain how the marketplace, with its "invisible hand," not only is the most rational and efficient means by which individuals can pursue their own interests, but is also the best means to raise the general standard of living of everyone participating. That is, while it brings different degrees of prosperity to each, it tends to bring the greatest degree of prosperity to all. When they address altruism, it is turned into a form of enlightened self-interest or reciprocal altruism. For example, if one pays taxes, it is not as a gift to others, but as a way of increasing the attractiveness of the environment in which the taxpayer lives, the value of which can then rebound to that particular taxpayer.[22] The most intense connection between self-interest and capitalism lies, as we shall see, in its conception of value. This is a technical idea, but its core is that value depends on the amount of pleasure, satisfaction, or "utiles" one receives in consuming a product. This commitment to self-interest can be loosened, and there have recently been calls among economists to do so, but not, it appears, through conventional economic theories.[23]

The theories of socialism/communism appear to pay greater attention to altruism. This is found in their emphasis on fraternity and community needs; the concern with equality of income; the role, in Kropotkin's terms, of mutual aid and cooperation that pushes us in the direction of cooperation rather than competition; and most famously, the arguments of Marx that communism will be based on the idea that we will and should create and produce not only for ourselves, but for the sake of others.[24]

The waters though soon become muddy, for in many of these socialist positions one also finds a sense that mutual aid and cooperation, care for others, communal benefits, equality, and the communal ownership of the means of production are motivated not as much by altruism as by reciprocal altruism or enlightened self-interest. That is, the individual finds it in his or her interest to cooperate rather than compete. And even if there really is true altruism within a community, it may not exist beyond its borders. Cooperatives, communities, associations, and states, though perhaps altruistically motivated within, can still be motivated by a corporate sense of self-interest when it comes to relations between and among them.

An additional difficulty in keeping altruism at the core even within a community occurs when newer Marxist theories introduce a market mechanism to provide some degree of coordination of interests and contributions. This raises again the basic question of how a system of altruism can be part of a market economy that is designed to reflect and quantify interests. How is self-interest kept at a minimum in such a schema? Here, the discussions of market socialism seems to be disappointing, for the altruism aspect is all but forgotten in discussions of efficiency.[25] Even so, it seems that socialism still allows greater potential for altruism than does capitalism. Both though can in principle accommodate it. The deeper question then is: How can altruism become the primary value undergirding these or other economic systems and, specifically, how can altruistic gift-value become the system's primary concept of value?

Theories of Value

Every culture recognizes, creates, circulates, and consumes things that it finds important. These things of value can be thought of as the economic foundation of the culture. Perhaps the most general concept of value is simply that a thing has value if it allows us to attain a goal. This can be thought of as value-in-use, or use-value. (I do not want to call this instrumental value, because the goal can be altruistic and that of doing good. So here, I will simply follow the convention of calling it use-value.) Food is useful because it satisfies hunger and nourishes our bodies. A house is useful because it provides shelter. And books are useful for the knowledge they contain. In nonmonetary economic systems (e.g., simple hunting and gathering societies) use-value would be the self-evident and the only concept of value.

The meaning of use-value is also connected to need. What is of use is that which satisfies a need. What then do we know about needs as a foundation of value? Are there more or less basic needs and are some really "wants" rather then "needs?"

Needs and value   The point here is that without a concept of the good, the search for more or less basic needs will not succeed. We might ask, for example, if the most basic human needs are biological, social, or intellectual. Elements from any one of these empirical realms can at one moment or another become essential, but there is no stable relationship because our needs depend on our contexts and our goals. Just as we can be biologically starved, we can be starved socially or intellectually. Just as one can suffer excruciatingly from not having enough to eat and drink, the same degree of pain can result from social ostracism and from intellectual anguish. Each can be fatal. This is why such contemporary attempts like that of Maslow's to develop a hierarchy of needs—where at the bases of his model are physiological (or biological) needs, and then in ascending order are security, social needs, esteem, and self actualization—have not held up to the evidence.[26] Indeed, research into basic and less basic needs and need hierarchy has virtually ceased. And when one takes a broader historical-geographical view on the matter, it is clear how different realms shift in importance. Currently the biological appears to be offered as the primary one, but in the Middle Ages, our bodies and physical appetites were thought to be the least important, for they distracted us from the mental and spiritual, wherein the most basic human needs were thought to lie.

This difficulty in defining needs is not a surprise if the issue is framed geographically. Place and self interthread and hybridize elements of meaning, nature, and social relations. While overall each realm is equally important, one may become more dominant in a particular situation. Because these situations are contingent, we can expect to see a complex and shifting set of needs emerging and receding, being created and discarded. Our selves and our places can never be completely perfectly matched. We cannot expect to be always in the right place at the right time, or for a particular place to work perfectly. This lack of correspondence will always set in motion one or more of the elements of nature, or meaning, or social relations, as something that needs attention. Needs not only shift, they also become blurred with "wants"; a condition that is exacerbated when the entire socio-geographic system, through the marketplace, instills in consumers an ever-expanding appetite for needs and then sells the consumer ever more products to satisfy this appetite. The economy grows by creating, satisfying, and then creating more wants or desires and having them appear as needs.

An important step in moving away from the purely instrumental view of needs is to follow Amartya Sen's suggestion and think of them not as ends needing to be satisfied, but as means to increasing the capabilities of humans to lead the lives they value "and have reason to value."[27] A society's wealth then is based on all of those things that it can offer to help to increase human capabilities, and these things can change geographically and historically. This is a great improvement, but can still revert to the instrumental if we do not take one more step and make clear that the reason to value certain kinds of capabilities is that they promote the good. Things are of value when they meet our needs to collectively see through to the real and to increase variety and complexity. As this becomes accepted, the economy will then be generating a form of wealth that can be called gift-value. This concept is best understood if it is seen in contrast to another meaning of value—that of exchange value.

Exchange value Things that are valued because of their use can also be exchanged. Exchange can be simple, as when "x" is swapped for "y," or it can be complex and general, as when money becomes used as a common means of exchange. Instead of swapping "x" for "y," I can now sell "x" for a certain quantity of money, and use that to buy "y," or anything else of equivalent value. But this idea of equivalent value is complicated when "x" and "y" are assigned prices, for it means that something is of value not only to me, but also to others who are involved in the process of exchange. The problem becomes central to an economy when things are produced primarily as commodities—as things intended to be bought, sold, and thus exchanged. So how then do we arrive at value—an exchange-value—for such things, or how do such things come to have a price?

Two kinds of answers have been provided: one is based on Marxist analysis, and the other on conventional economic theory. Both allow exchange value to be quantified into a price. The Marxist view would argue that exchange value reflects the amount of human labor embodied in the commodity. Exchange among commodities is possible because underlying all commodity production is a common or homogenous quality that stems from the fact that they are all products of human labor. Different types of labor are involved in the production of each type of commodity, but by exchanging them, labor is in effect being generalized and abstracted. Exchange-value (or price) then is the quantitative form this abstract labor takes. The idea that labor is the basis of value—a *labor theory of value*—is controversial even within Marxism, and there are several variants of it, but the important and general point is that laboring involves complex social relations (and in capitalism it involves, according to Marx, the all-important one of exploitation of workers by capitalists) and so exchange-value and money are ultimately products of labor as it is molded by these socioeconomic relations.

The second answer comes from conventional economic theory. There, exchange-value becomes more a psychological than a social concept. Exchange-value is not linked to labor but rather depends on the relationship between the psychological satisfaction (or pleasure or utility) that a commodity could provide and its availability. This relationship can be phrased more generally and for a marketplace as the connection between supply and demand. (In this view, the exchange-value of a commodity is equivalent to marginal utility of that commodity, and use value to total utility.)[28]

It is not important here to argue the relative merits of these two very different theories of exchange-value, but rather to quickly point out that they do have things in common that will serve as a contrast to our discussion of gift-value. Both theories can be used to explain how the accumulation of money can appear to be an end in itself. They both would argue that for a capitalist system to generate wealth it must constantly expand, and commodities and money must constantly circulate. Both would agree that, instead of being driven by specific needs that could be satisfied by particular commodities, the system encourages the constant expansion of needs with ever more commodities to fill them, thereby creating the ever increasing need for money. Individuals are practically forced to see the abstract accumulation of money as either an end in itself, or as the universal and indispensable means to all other ends.

As this system of exchange expands into other realms of life, making them part of the economy, it also brings with it an instrumental quality to social relationships that is based on enlightened self-interest[29] that can be called "businesslike"—a tepid form of social relations. This business relationship with its exchange of goods and circulation of money can cool off close friendships and family ties by making them based on a calculated form of mutuality; this same business relationship can also serve to build a degree of trust among those who are perfect strangers and even enemies.[30] But the relationship is instrumental: the motivation for the social bond is temporary and provisional, being ceaselessly redefined by changing social relations.

Circulation is crucial, but withholding things temporarily from circulation—including meaning and knowledge, material objects, and even social relations—can often increase their value. And, to the extent that individuals are accumulating things under capitalism, they are also withholding them from the public. The private domain expands, while the public contracts. On all of these counts both theories agree. And they agree too that capitalism must have some kind of a public realm to provide for elementary public goods and to establish a political order, but there will also be strong incentives for the public realm to be as small as possible.

The central point is that both economic theories see exchange-value as a quantitative relationship that is almost always driven by a psychology of self-interest. For the committed supporters of market capitalism, this is a good thing for they see self-interest and the market leading to virtuous outcomes. Sympathetic critics, though wanting to blunt the focus on self-interest, think they can do so within the overall context of exchange-value by building into the exchange system specific quantifiable incentives to encourage giving and caring. This can be accomplished when it is pointed out, for example, that mutual cooperation produces positive externalities and network effects that lead to increasing returns up to a point;[31] or it can be encouraged by introducing schemes that substitute higher status and recognition rather than higher wages and salaries for those who are more productive and who do socially more desirable things.[32] There are a few who attempt to make altruism coequal to self-interest in the logic of exchange. Kolm for example argues that the same concepts of utility functions, preferences, supply and demand, and market efficiency can apply to a system that is based either on altruism or self interest.[33] And again, there are those who reject capitalism and promote a market socialism instead.[34]

These works may offer the most effective and convincing means of showing that a market can allow us to be less self-interested, and thus induce us to become so. My problem though is that they still use self-interest as the broader context and have altruism make sense within it. It is as though intrinsic judgments are, in the last instance, justified by instrumental ones. Again, the hold that self-interest has on economic theory in general, and on theories of value in particular, makes it difficult to see what else can be done. So far, attempts to modify or move away from self-interest become enfolded back within the scope of self-interest. The problem is that we are trying to loosen the grip of self-interest by backing away from it. Perhaps we should turn around and simply face in another direction.

Intrinsic judgments can help. They remind us that we have the capacity to decide what is important and what we ought to do. And this applies also to what is of value. If we ought to be good, and if altruism is central part of being good, then we ought to have a system of values that privileges this. The qualities of altruism that are akin to the economic are its gift-values. Gift-value has properties that are different from exchange-value. One of them is that gift-value is not readily quantifiable and subject to marginal utility. Even though there are advantages to considering things as though they were quantities,[35] I do not think quantification does sufficient justice to the real power of gift-giving. It is worth considering how this power differs from exchange-value and how it works on its own. To do so may clarify why we would want altruism to be a fundamental property of our economic models, and may open up the possibility of altruism animating the value of exchange, thereby making the instrumental more like the intrinsic.

Before I do so, I must point out that to have gift-value make sense, it must "speak" the language of abstract economics. This means I must continue to express its qualities apart from the more integrative terms of place and place-making. But this does not mean that geography is not informing the discussion. Quite the contrary. It is intrinsic judgments that draw attention to the very concept of gift-value, and it is the theory of place that points to the fact that gift-value can be created and circulated in any one (or combination of) the realms. These ideas will structure the discussion. But the full force of the geographic occurs when we recognize that the concept of gift-value is fully realized only when it is recombined with the other components of place and the moral theory.

Gift-value How then does gift-value differ from exchange-value? As with exchange-value, gift-value requires constant circulation. But an extremely important and obvious difference is that the circulation is motivated by giving rather than exchanging and accumulating. The circulation and value of the gift depend on it not only being given away, but on having the person or persons receiving it also be motivated to give things away. The gifts, as they circulate, become transformed; they inspire other and different generous and kind acts. If the recipients are not inspired, and do not in turn sooner or later give, the value of the gift stops. It becomes instead another type of use-value, or something to be exchanged. It remains a gift-value only if it is free and freely given. Gift-value exists only as long as there is this "inspirational" effect that is linked to its circulation.[36]

This important connection between inspiration and circulation is true in each of the realms within which gifts circulate. The gift-value of a Beethoven symphony depends on its availability to others who may read the score, play in a symphony, or hear the music. Each person who has access can benefit. Each time it circulates it may change its meaning and import, and inspire and be incorporated in other musical creations that then, in turn, circulate as contributions in different contexts and places. Its capacity to expand our horizons increases, as there is more of it and as more is added to it. Or, consider Galileo's contribution to the heliocentric view of the world, which he imparted at great personal risk. Newton's work drew on Galileo's, and Einstein's on Newton's. As it became incorporated within a new form and circulated, its meaning changed, and further expanded our horizons. So, a gift's form does not remain constant and its value increases as it becomes used and incorporated into new gifts. In the same way, the gifts of care and justice are part of the system of gift-value insofar as they inspire further examples and circulation. The same is the case with the natural when it reminds us of the plenitudinous givenness of reality. The more available a gift, the more each instance of its gift-value increases.[37]

Another important difference concerns the quantitative relations of exchange, and the qualitative relations of the gift. Exchange-values, as they are discussed in economics, are entirely quantitative. Not only are different types of things comparable in terms of their price, but under market conditions, commodities tend to be mass-produced, so that there are many instances of the same product: there are hundreds of thousands of a particular make and model of car manufactured, and there are millions of identical cans of chicken soup sold each year. But the core meaning of gift-value is difficult to put into numbers. This distinction does not rest on the fact that many gifts, such as knowledge and care, are often less or nonmaterial, while other commodities are more so. Money after all is not essentially material, but is certainly quantified and at the basis of exchange-value; and acts of charity may involve material things that have measure and physical substance, but as gifts their importance may not. True, we often say that some gifts have been more significant and possess more gift-value than others; still, we cannot in any reliable sense measure this significance or value. The same is true when we say that "this" is more beautiful than "that," or that the beauty of "this" has increased. Though we think these terms about "increasing" and "decreasing," and "more" or "less," signify real relationships, we do not expect them ever to be quantified.

There are several reasons for the unsuitability of quantification. Perhaps the most important rests directly on the inspirational effect of the gift, which takes time and grows over time. A gift's significance may not be determinable immediately even for a single recipient. I may not know how important a particular act of kindness has been to me until I realize how it has inspired me to be kind. And then the value of my kindness increases as others become inspired. Another is that gift-giving tends to focus on the uniqueness of the gift rather than on how it may be an instance of a class or group. This is because gift-value depends on the gift inspiring other gifts. An idea that was taught to me, though again an instance of a type of thing, is still different from all other ideas in how it affected me. How it is used, absorbed, and contributes to my own context makes the differences seem important.

A more general way of saying this is that gift-value does not seem to be subject to the law of supply and demand. That is, gift-value does not follow the exchange-value logic that as supply increases, the price or value of the object decreases, or to put it another way, that the law of diminishing marginal utility does not apply to gift-value. Rather, it may tend to operate in the opposite way. Increasing supply can increase the value of the gift. The reason gift-value does not follow declining marginal utility is its inspirational effect. This makes it difficult to imagine overall that there can be too many acts of kindness or goodness or instances of justice, or too much truth and beauty. For if there were, we would be saying that the real and the good themselves are neither infinite nor compelling. Or, to put it another way, we would be claiming that we already know enough about the world, and have enough acts of kindness, or enough goodness, thank you!

In contrast to gift-value, it is possible (though economically suicidal from the viewpoint of exchange-value and capitalism) to say that there is too much wealth in a country, or that the economy should not grow, or that we can do without more commodities. This after all is the cry of many who find the continuous accumulation of material goods and wealth to be empty, and who seek other economic systems. I am not saying that theirs is a practical view, but that it is, for many, a morally appealing alternative to a system that is based on growth for growth's sake, and that ultimately may not be sustainable. By the same token it is difficult to imagine that a gift-giving system should be limited in the same sense of limiting one based on accumulation; for again, there can never be too much goodness.

It may seem that gift-value has much in common with public goods—goods that, even in a market system, must be supplied publicly and used publicly. While public goods can, in principle, be incorporated under the concept of gift-value, they are usually conceived of and supplied in a way that makes this difficult if not impossible. For example, public goods are provided by a collectivity, whereas a gift is often also given by an individual; and in public good theory, a public good usually requires some sort of forced contributions by the public—a tax for example—in order to avoid the "free rider" problem, while altruistic gift-giving is not forced or taxed, and free riders do not present a problem.

While there does not seem to be diminishing marginal utility to gift-giving at the general or macro level, at the individual or micro level the issue may be murkier. At the receiving end, a person may feel smothered with gifts at any particular moment. This is obviously the case for a hungry homeless person who is presented a thousand cans of chicken soup. In this case diminishing marginal utility is the obvious mechanism to explain the person's response to all of these cans. But it can also happen more subtly with immaterial gifts. I may feel, at a particular moment, that I can no longer absorb another fact or bit of wisdom, or stand to take another act of kindness without being smothered. In these cases, it seems that my personal reaction to receiving or consuming goodness operates in the same way that personal preferences for commodities operate in conventional exchange-value models. Indeed, we have only a limited capacity to be aware of reality and the good; we can only take so much before we need to escape. In addition to something like diminishing marginal utility, gift-value may be following the principles of opportunity costs. An intellectual gift may be best used by having us think about it, mull it over, and "digest" it over a long period of time, during which we then close our minds to other similar types of gifts. Those undertaking the project cannot then undertake another one, and those receiving and digesting a gift lose the opportunity to receive and digest another.

From one direction, these points ring true, but from another they don't, for it does not seem right to say that we should at any moment be content with our access to truth, justice and care, and beauty, or that we do not want to help make a world that is richer in its variety and complexity and capacities to help us see more clearly. And while we may in fact feel "full" at the moment, it is a different kind of fullness than comes from consuming commodities in a system of exchange. It is more like the fullness that comes because of the time and even effort it takes to appreciate what is given and to know how best to continue the circulation of gift-giving, and it comes also with a sense that our capacity for appreciating instances of the good should be expanded. I should want to know more and to see more beauty, and I should feel guilty when I have reached a temporary limit, because my reception of these gifts maintains them as gifts only if they in turn inspire me to give. And while it is true that when creating or receiving a gift I am then not engaged in other projects and opportunities and so am incurring opportunity costs, these are experienced differently from opportunity costs in exchange-value, for gifts are made public and so others and I will not ultimately lose them.

As an ideal then, gift-value suggests an open-ended spiral of ever-greater value that is accessible to anyone. Gift-value provides "cultural capital" for all (though I avoid the phrase because it comes from a literature that privileges social relations and theories of self-interest and rational choice).[38] Of course this type of value must be connected to other kinds, especially exchange-value. Again, all I can do is suggest that this may not be insurmountable if we remember that value itself is what we decide it to be, and so if we accept altruism and altruistic gift-giving as a model, this can then animate those who produce and consume in an exchange or market system. Exchange-value becomes an appendage to gift-value, and not the other way around.

The full import of this is clear as we realize that gift-value is part of intrinsic judgments, that its qualities are ones we associate more generally with altruism, and that every aspect of this abstract discussion on value was designed to reinsert the economic back into the world— a world that contains the political and the social as well as the realms of meaning and nature and the moral and aesthetic domains: a world that comes together by place-making. But to say this is no longer to "speak" only the language of economics, but rather to have economics become informed by geography and even by intrinsic judgments, if we accept them. How the economic becomes entwined with, and altered by, all of the other parts of place can be suggested by one ordinary example. Consider again a restaurant that goes out of its way to inform its customers about its labor practices, keeps its books open for public inspection, encourages its workers to be involved in decision-making, explains where its "raw" materials come from, and offers food that is affordable, enjoyable, nutritious, and in an aesthetically pleasing environment—in short, a place that is guided by intrinsic judgments. In this case, the restaurant's price for its meals (its exchange-value) will be informed by gift-value. This place can have an inspirational effect. If another restaurant were to open that engaged in similar practices, this would not then force this one to pull back from these measures, but inspire it to continue to move even farther in this direction, and to expand these practices in a different way. This kind of an example fits what many economic and environment reformers advocate in their appeals to openness and democratic process. It is in short an example of exchange-value animated by gift-value, and its likelihood depends on what we decide to value.

I have stressed gift-value and gift-giving. What though of those who receive a gift? Is there an inequality in this relationship that leads to difficulties, such as a sense of superiority and power on the part of the gift-giver and on the recipient a feeling of resentment, a diminution of self-esteem, and even envy? The theory would not encourage these because they are emotions associated with self-interest and not with gift-giving. A Beethoven or Einstein would no doubt prefer recognition and gratitude, but their gifts did not seem to depend on it. And those who have received them, should not thereby feel less adequate, humiliated, or gripped by envy; rather they should feel inspired. The spirit of gift-value is generosity, and when we think of generosity, and the associated qualities of justice, and care in the context of intrinsic geographic judgments, we see that there are innumerable ways of giving gifts. Creating places that are varied and complex and that allow us to see reality more clearly encourages multiple ways to contribute.

This is phrased in terms of individuals, but does it work with groups and even whole cultures? I see no reason why not. A collectivity of individuals engaged in a practice, such as a business, a school, a community and even a culture, can find inspiration in the practices of others as readily as can a single individual. Cultural borrowing and giving is a good thing, and is the counterpart of having a place and culture open to immigrants who then are willing to become engaged in the place's practices and to participate in their changes. Altruistic gift-giving assumes that people and ideas should circulate under the guidance of intrinsic judgments. If these conditions are not met, then such movement will likely serve the interests of one party and lead to a tyranny and homogenization of culture and ideas. To preserve and protect a culture from giving and receiving gifts is to keep its members from seeing clearly and making choices. Keeping territorial boundaries of a culture open will no doubt lead to the complete transformation and even abandonment of some ways of life and to the creation of new and different ones to replace them. This is part of the geographic condition, and it does not have to be directionless or pointless if there is the possibility of a guide to make sense of what these processes mean.

Having an economy inspired by altruism is a move in the right direction. But altruism must also animate our political and social institutions and these must be connected to the economic. How then are these related? This again is where place enters and our place-making becomes central. In a specific sense, the structure and dynamics of place explain how these and other facets of our experience are related: it is through the structure and dynamics of place that we can act simultaneously as intellectual, social, and natural beings, and still specialize in one of these here and another there. Place enables these connections and undertakings, and is also the means by which they are circulated and received. In a more general sense, place is central because we cannot consistently adopt an altruistic view and have it infuse our actions without employing intrinsic judgments in our place-making activities. The claim of the theory is not only that altruism is attractive, but that it has a chance of being enacted only if our places are guided by intrinsic judgments. The theory and the problematic remind us that our geographical nature uses our will and imagination to constantly change reality and intrinsic judgments urge us to change reality under the guidance of the real and the good. As these judgments are taken up and direct our place-making, they help us interconnect the threads woven by place—including our economic, political, and social activities. If our place-making is animated by moral concerns, then these will necessarily become part of our projects.

Good social institutions can assist us in pursuing the good; but they neither force us to seek it, nor determine it. (This is why our discussion of social relations assumed that we have already accepted intrinsic judgments.) The theory then puts the moral argument and the role of free will ahead of the empirical process and structures. It sees the power of the moral as transformative of the empirical. It also recognizes that, apart from general suggestions such as having institutions with open and flexible territorial boundaries, free exit and entry, intrinsic democracy, and an economics informed by gift-value, we cannot specify precisely what good social institutions would be. This will depend on our own opportunity and context and on our geographic imaginations.

It shows also that doing good is rarely (and then mostly desperately) dependent on a single heroic act. Rather, good (or bad) actions usually involve a series of countless small steps. This means that we can make modest but significant progress in the most practical and mundane activities. Intrinsic judgments would urge that any undertaking, whether it be a grand political system or the running of an office or restaurant, involve the continuous efforts by its members to reflect on and discuss how their roles help to contribute to heightened awareness and variety and complexity. Not only must the product of the undertaking be a contribution, but also the process of production must enlighten those engaged in it. The structure and dynamics of the place should be as open and transparent as possible, and the processes of giving should be expressed in care and attention to the role of this place in a network of places.

As a society avoids moral drift and is pointed in the right direction it also eliminates the need for heroic acts of self-sacrifice, since our daily actions will involve care and attention, and a heightening of awareness. A good society occurs only if we, with the assistance of good institutions, make a continuous effort to attend to these matters. The urgent need to attend is evident as soon as we remember that at the very moment we are attempting to create places and institutions that help increase our awareness of reality and the variety and complexity of reality, we encounter place's frictional effects of opacity and compartmentalization that can all too easily lead us back to moral drift and worse.

Postscript

8

THE PROBLEMATIC AND MORAL THEORY

WE ARE PLACE-MAKERS WHO transform the earth. This is our geographic condition, and understanding it sheds light on who we are as moral agents. As framed by the problematic, it becomes the process of not being able to accept reality as it is and to continuously create places to transform it into what we think it ought to be. The *ought* is where the moral enters. The *ought* ought to be inspired by an intimation of the good. Even though the good is ultimately ineffable, facets of it can be made more accessible through intrinsic judgments which claim that we *ought* to create places that jointly increase our awareness of reality, and increase the variety and complexity of reality.

Intrinsic judgments serve as a compass bearing. They help guide how we transform the world and give our changes purpose and direction. They recognize that we cannot accept reality as it is because we *must* make it better. Having the purpose of making reality better opens up several lines of thought. It encourages us to think of ourselves as purposeful moral beings who have free will and the capacity to be altruistic. It points to the moral as a force affecting the empirical (rather than the other way around) and stresses the benefits of being less situated and self-interested. Its focus on free will and altruism interprets the empirical regularities disclosed by social science to be the result of our (intentionally or unintentionally) following rules and conventions of conduct that we can change, if we find it morally compelling to do so. It reminds us that the moral is not only a matter of justice, but also of truth and the natural. And its recognition of the reality but ineffability of the good encourages us to seek it with the understanding that any sense of it would be partial and provisional.

The theory does not disclose new moral facts: the value of seeing the world more clearly and of living in a complex and varied world are known. Rather it brings these together in a way that can explain how geography—a field that is usually associated with the literal and concrete—has a bearing on abstract and ineffable matters such as the real and the good, and how these are woven together and contextualized, though not relativized, in our everyday place-making at each and every geographical scale.

We will forever be engaged in transforming the Earth. Our motivation for doing so is that we have conceptions of what we think the world ought to be. Many of these conceptions are self-centered and tragically misdirected, but we can know this and correct it only through an intimation of the good. The theory attempts to put that intimation more explicitly and geographically. By encouraging us to create places that increase our awareness of reality and the variety and complexity of that reality, the theory can sharpen our critique of bad places, provide suggestions for improvement, and offer a positive image of what good places could be like. It helps us navigate between the equally dangerous pitfalls of moral absolutism and moral relativism, and it clarifies what moral progress means. In all of these ways, it provides us a reason for hope.

NOTES

Chapter 1

1. The geographic problematic changes and draws from a formulation developed by Yi-Fu Tuan regarding culture as a means of escaping from reality. In his *Escapism*, Baltimore: The Johns Hopkins University Press, 1998, p. 6, he says: "A human being is an animal who is congenitally indisposed to accept reality as it is. Humans not only submit and adapt, as all animals do, they transform in accordance with a preconceived plan. That is, before transforming, they do something extraordinary, namely, "see" what is not there. Seeing what is not there lies at the foundation of all human culture." I have shifted the emphasis from escaping to transforming and have made it more geographical by stressing the role of place.

2. My definition of place incorporates the idea of place-as-territory, simply territoriality. See R. D. Sack, "Author's Response," *Progress in Human Geography*, 24(1), 2000, pp. 96–99, for a discussion of the relationship between place as defined here, and my definition of human territoriality in Human Territoriality: Its Theory and History, Cambridge: Cambridge University Press, 1986. There are many other important definitions of place that differ from mine. See, for example, Richard Hartshorne, The Nature of Geography: A Critical Survey of Current Thought in Light of the Past, Lancaster, PA: Association of American Geographers, 1939; Yi-Fu Tuan, Space and Place: The Perspective of Experience, Minneapolis: University of Minnesota Press, 1977; J. Nicholas Entrikin, The Betweenness of Place: Toward a Geography of Modernity, Baltimore: The Johns Hopkins University Press, 1991.

3. See, for example, David Harvey, Social Justice and the City, Baltimore: The Johns Hopkins University Press, 1973 and Spaces of Hope, Berkeley: University of California Press, 2000; David Smith, Moral Geographies: Ethics in a World of Difference, Edinburgh: Edinburgh University Press, 2000; Yi-Fu Tuan, Morality and Imagination: Paradoxes of Progress, Madison, WI: University of Wisconsin Press, 1989; James Proctor and David Smith, eds. Geography and Ethics: Journeys in a Moral Terrain, London: Routledge, 1999; and the journals on geography and ethics and philosophy and geography.

4. I am saying no more here than, as Colin McGinn puts it: "what is wanted, then, is a philosophy of morality that shows it to be an area of truth and knowledge in good standing, despite the absence of the divine authority that has traditionally been supposed to give it substance." Colin McGinn, Ethics, Evil, and Fiction, Oxford: Claren-

don Press, 1997, p. vii. Notice that I sometimes use the terms morality and ethics interchangeably, but more often I stick to morality because I think it is the broader of the two, and subsumes ethics.

5. This is a Kantian view and assumes an autonomy on the part of a moral agent and leads, in Nagel's terms, to what may seem like an "uncaused" cause (What Does It All Mean?, Oxford: Oxford University Press, 1987, p. 50). The moral realm works with reasons instead of causes, and our will allows us to accept or reject them.

6. Moral naturalism is discussed by Charles Pigden, "Naturalism" in Peter Singer, ed. A Companion to Ethics, Oxford: Blackwell, 1991, pp. 421–430. See also the article by Michael Smith on Realism, pp. 399–409, in the same volume.

7. One can say that this is a critical or skeptical realism. For an overview of realism see Frank B. Farrell, Subjectivity, Realism, and Postmodernism, Cambridge: Cambridge University Press, 1994, and also Farrell, "Rorty and the Antirealism" in Herman Saatkamp, ed. Rorty and Pragmatism, Nashville: Vanderbilt University Press, 1995, pp. 154–188, where he proposes a definition similar to mine.

8. Colin McGinn, Problems in Philosophy: The Limits of Inquiry, Oxford: Blackwell, 1993, p. 5, accepts a strong form of realism as he puts it, which is what I call critical realism.

9. The opposite view is expressed by John Kekes in Facing Evil, Princeton: Princeton University Press, 1990. Seeing little room for free will, his is a more conservative view that tries to keep such "evil doers" apart from the rest of the population, and does not place much store in the remedial role of education.

10. Peter Kropotkin, "What Geography Ought to Be," The Nineteenth Century, 18, 1885, pp. 940–956, excerpted in John Agnew et. al, eds. Human Geography: An Essential Anthology, Cambridge, MA: Blackwell, 1996, pp. 141–142.

11. James Lovelock, The Ages of Gaia: A Biography of Our Living Earth, New York: Norton, 1995, pp. 97–98.

12. Daniel Goldhagen, Hitler's Willing Executioners: Ordinary Germans and the Holocaust, New York: Vintage Books, 1997, p. 458.

13. Instrumental and intrinsic will be the preferred terms because pure and practical may be confused with Kant's use of these terms in his Critiques (Kant, 1952).

14. Though I am aware of the arguments concerning the tyranny of the gaze, I use the word "see" to connote "knowledge" and "understanding" because it is apposite to geography.

15. Here I am collapsing several human qualities that others have dealt with separately. Consider Stephen Holmes, "The Secret History of Self-Interest" in Jane Mansbridge, ed. Beyond Self-Interest, Chicago: University of Chicago Press, 1990, pp. 267–286; Albert Hirschman, The Passions and the Interests: Political Arguments for Capitalism Before Its Triumph, Princeton, 1997; and David Hume, A Treatise of Human Nature, Oxford University Press, 1888.

16. For general discussions of altruism see Thomas Nagel, The Possibility of Altruism, Princeton: Princeton University Press, 1970, and for examples of altruistic behavior see Kristen Monroe, The Heart of Altruism: Perception of a Common Humanity, Princeton: Princeton University Press, 1996. On reciprocal altruism see Lawrence Becker, Reciprocity, London, Routledge & Kegan Paul, 1986.

17. This may capture the spirit of Marx's Comments on James Mill, pp. 227–228.

18. Marcel Mauss, The Gift: Forms and Reason for Exchange in Archaic Societies, New York: Norton, 1990.

19. Iris Murdoch attempts to write novels whose protagonists are good characters.

20. For a geographic critique see, for example, Robert Sack, Place Modernity and the Consumer's World. Baltimore: The Johns Hopkins University Press, 1992.

21. For Auschwitz see Dominick LaCapra, History and Memory After Auschwitz, Ithaca: Cornell, 1998, who cites literature on monuments to horrors; for the relationship of

place to memory see Ken Foote, Shadowed Ground: America's Landscapes of Violence and Tragedy, Austin: University of Texas Press, 1997; Steve Hoelscher, Heritage on Stage: The Invention of Ethnic Place in America's Little Switzerland, Madison: University of Wisconsin Press, 1998; Karen Till, "Reimaging National Identity" in Paul Adams et. al. ed. Textures of Place, Minneapolis: Univeristy of Minnesota Press, 2001, pp. 273–299; and David Lowenthal, The Past Is a Foreign Country, Cambridge: Cambridge University Press, 1985.

Chapter 2

1. Robert Sack, Homo Geographicus: A Framework for Action, Awareness, and Moral Concern, Baltimore: The Johns Hopkins University Press, 1997, develops the theory in detail.

2. Sack, Op. cit.

3. Ronald Dworkin, Sovereign Virtue: The Theory and Practice of Equality: Cambridge, MA: Harvard University Press, 2000, attempts to find a common thread underlying the various facets of justice and proposes a form of equality as the solution.

4. Other ways of characterizing aspects of truth are found in Felipe Fernandez-Armesto, Truth: A History and a Guide for the Perplexed, New York: St. Martin's Press, 1997.

5. Bertrand Russell, Human Knowledge: Its Scope and Limits, New York: Simon and Schuster, 1962, pp. 60–61.

6. There are many discussions of the types of truth. See, for example, Carl Hempel "On the Logical Positivist Theory of Truth" Analysis, 2, 1935, pp. 49–54 and Hilary Putnam, Reason, Truth, and History, Cambridge: Cambridge University Press, 1981, for a coherence theory; John Searle, The Construction of Social Reality, New York: Free Press, 1995, for a correspondence theory. Jurgen Habermas's work comes close to espousing a consensus theory and, of course, any fundamentalist religion argues for a revelatory theory.

7. Tim Cresswell, In Place/Out of Place: Geography, Ideology, and Transgression, Minneapolis: University of Minnesota Press, 1996, has an excellent discussion of the in/out rule.

8. This is the purpose of the older spatial analysis models, which addresses the role of space and distance in the models of spatial and locational analysis. See Richard Chorley and Peter Haggett, eds. Models in Geography, London: Methuen and Co. Ltd., 1967. For a newer image of spatial flows and interactions, see Manuel Castells, The Informational City, Oxford: Blackwell, 1989, and Manuel Castells, The Rise of the Network Society, Cambridge: Blackwell, 1996.

9. Don Mitchell raised this point in a 1999 lecture at Madison, Wisconsin.

10. Edward Soja, Thirdspace: Journeys to Los Angeles and Other Real-and-Imagined Places, Oxford: Blackwell, 1996, employs this term, as well as trialectic.

11. Manuel Castells, The Informational City, Oxford: Blackwell, 1989, and The Rise of the Network Society, Oxford: Blackwell, 1996.

12. See Neil Brenner, "Limits to Scale? Methodological Reflections on Scalar Structuration," Progress in Human Geography, 25(4), Dec. 2001, pp. 591–614.

13. Donna Haraway, Simians, Cyborgs, and Women: The Reinvention of Nature, London: Free Association Books, 1992, and Bruno Latour, We Have Never Been Modern, Hemel Hempstead: Harverster Hewatsheaf, 1993, discuss the hybridization of nature and culture. But they do not provide a mechanism like place to weave these components. See also Jonathan Murdoch, "Inhuman/Nonhuman/Human: Actor-Network Theory and the Prospects for a Nondualistic and Symmetrical Perspective on Nature and Society." Environment and Planning D; Society and Space, 1997, pp. 731–756, who

reviews these issues, but again does not offer an understanding of how geography plays center stage. To use Harroway and Latour as the ultimate authorities on this matter is to run the risk of ignoring the foundational role of geography.

14. Cresswell, op. cit., provides excellent examples of transgressions.
15. The structuration theory of Anthony Giddens, The Constitution of Society, Berkeley: University of California Press, 1984, sidesteps the issue.
16. John Searle, Rationality in Action, Cambridge, MA: MIT Press, 2001, makes a similar point. See Dave Delaney, "Making Nature/Marking Humans" Annals, AAG 91, 2001, pp. 487–503, who points out how difficult it is to isolate it.
17. Derrida has written a vast amount. Simon Critchley, The Ethics of Deconstruction: Derrida and Levinas, West Lafayette, IN: Purdue University Press, 1992, attempts to cull moral implications from this vast work.
18. Michel Foucault's works are extensive. An excellent review of the phantom role of morality in his work is Nancy Fraser, Unruly Practices: Power, Discourse and Gender in Contemporary Social Theory, Minneapolis: University of Minnesota Press, 1989, pp. 1–35.

Chapter 3

1. Alasdair MacIntyre, After Virtue: A Study in Moral Theory, Notre Dame, IN: University of Notre Dame Press, 1981, p. 11.
2. Overviews of communitarianism are found in Stephen Mulhall and Adam Swift, Liberals and Communitarians, Oxford: Blackwell, 1992, and Shlomo Avineri and Avner de-Shalit, Communitarianism and Individualism, Oxford: Oxford University Press, 1992.
3. MacIntyre, op. cit., p. 175.
4. Ibid, p. 178.
5. Michael Walzer, Spheres of Justice, New York: Basic Books, 1983, p. 6.
6. Ibid, p. 9.
7. Bernard Williams, Ethics and the Limits of Philosophy, Cambridge: Harvard University Press, 1985, p. 110.
8. Ibid, p. 111.
9. Walzer, op. cit., p. xiv.
10. Ibid, p. xiv.
11. Michael Walzer, Thick and Thin: Moral Arguments at Home and Abroad, Notre Dame, IN: University of Notre Dame Press, 1994.
12. See Paul Gross and Norman Levitt, Higher Superstitions: The Academic Left and Its Quarrels with Science, Baltimore: The Johns Hopkins University Press, 1994, and Ernest Gellner, Postmodernism, Reason, and Religion, London: Routledge, 1992.
13. Among the first to cull these points from Foucault's work was Nancy Fraser, Unruly Practices.
14. Op. cit., footnote 13.
15. See Simon Critchley, The Ethics of Deconstruction.
16. I owe these observations to Richard Kleinheinz.
17. Consider the postmodern interest in space over place to be exemplified by Gilles Deleuze and Felix Guattari, A Thousand Plateaus: Capitalism and Schizophrenia, Minneapolis: University of Minnesota Press, 1993.
18. I take this sense of the good by reading between the lines of writers such as Levinas, Derrida, Foucault, and their followers, such as Bauman and Critchley, and the geographic spin placed on these views by radical democracy and their discussion of public place. See Chapter 7.

19. "In a communist society there are no painters, but only people who engage in painting among other activities." Marx, The German Ideology, p. 394; as quoted in Jon Elster, Making Sense of Marx, Cambridge: Cambridge University Press, 1985, p. 81.
20. Jon Elster, Ibid., p. 87, quoting Marx's Comments on James Mill, pp. 227–228.

Chapter 4

1. Adolf Hitler, Mein Kampf (translated by Ralph Manheim), Boston: Houghton Mifflin Company, 1971, pp. 402–403.
2. John Lukacs, The Hitler of History, New York: Alfred Knopf, 1997, p. 182.
3. Mein Kampf. There allusions appear everywhere, but see p. 327 for an example.
4. Konnilyn Feig, Hitler's Death Camps: The Sanity of Madness, New York and London: Holmes and Meier, 1981, p. 11.
5. Feig, p. 11.
6. Gerald Fleming, Hitler and the Final Solution, Berkeley: University of California, 1984, pp. 186–188.
7. See, for example, Pierre Birnbaum and Ira Katznelson, eds. Paths of Emancipation: Jews, States, and Citizenship, Princeton: Princeton University Press, 1995.
8. George Mosse, The Crisis of German Ideology: Intellectual Origins of the Third Reich, New York: Howard Fertig, 1998, pp. 15–17.
9. Mark Bassin, "Race Contra Space: The Conflict Between German *Geopolitik* and National Socialism," *Political Geography Quarterly*, 6, 1987, pp. 115–134.
10. Deborah Dwork and Robert Jan van Pelt, Auschwitz: 1270 to the Present, New York: W.W. Norton, 1996, pp. 81–82.
11. Rudolf Höss, Steven Paskuly, ed. Death Dealer: The Memoirs of the SS Kommandant at Auschwitz, New York: Da Capo Press, 1996, p. 183.
12. Mein Kampf, pp. 57–58.
13. Ibid., p. 305.
14. Ibid., p. 120.
15. Ibid., p. 249.
16. Ibid., p. 303.
17. Ibid., p. 394.
18. "The Long and Bumpy Road of Jewish Emancipation in Germany" in Pierre Birnbaum and Ira Katznelson, eds. Paths of Emancipation: Jews, States, and Citizenship, Princeton, 1995, pp. 69–93 (cited in Goldhagen, p. 501).
19. Goldhagen, p. 56.
20. Goldhagen, p. 57.
21. Goldhagen, p. 58.
22. Lucy Dawidowicz, The War Against the Jews 1933–1945, New York: Holt, Rinehart & Winston 1975, p. 46.
23. Dawidowicz, p. 46.
24. Fleming, p. 17.
25. Hilberg, pp. 38–39. The Destruction of the European Jews, New York: Holmes and Meyer, 1985.
26. Janet Biehl and Peter Staudenmaier, Ecofascism: Lessons from the German Experience, San Francisco: AK Press, 1995, p. 15.
27. Ibid., p. 16.
28. Deborah Dwork and Robert Jan van Pelt, Auschwitz, p. 159. The words are Dwork's and van Pelt's summarizing Himmler's.
29. Ronnie Landau, The Nazi Holocaust, Chicago: Ivan R. Dee, 1992, p. 131.
30. Ibid., p. 132.
31. Ibid., p. 133.

32. Hilberg, p. 50.
33. Hilberg, p. 49.
34. In addition to Goldhagen's book, another well-documented argument for the complicity and German citizens in Hitler's plans is Eric Johnson, Nazi Terror: The Gestapo, Jews and Ordinary Germans, New York: Basic Books, 1999.
35. Landau, pp. 152–153.
36. Goldhagen, p. 145.
37. Hilberg, p. 338.
38. Hilberg, p. 338.
39. Goldhagen argues that efficiency was not the issue, rather it was to protect the killers from the slaughter, p. 157.
40. Landau, p. 316; Hilberg, p. 228.
41. Feig, for a discussion of the types of camps, and also for the number of deaths.
42. Wolfgang Sofsky, The Order of Terror: The Concentration Camp, Princeton: Princeton University Press, 1997, pp. 3–4. 1997, the original German version.
43. Goldhagen, p. 170.
44. Goldhagen, p. 157.
45. Eugen Kogon, The Theory and Practice of Hell: The Classic Account of the Nazi Concentration Camps Used as a Basis for the Nüremberg Investigations, New York: Berkeley Books, 1950, p. 44.
46. Ibid., p. 44.
47. Ibid., p. 42.
48. Otto Friedrich, The Kingdom of Auschwitz, New York: Harper Perennial, 1982, p. 49.
49. Sofsky, p. 71.
50. Friedrich, pp. 23–26.
51. Another purpose of the camps was as training grounds for the SS.
52. Mein Kampf, p. 414.
53. Claudia Koonz, Mothers in the Fatherland: Women, the Family and Nazi Politics, New York: St. Martin's Press, 1987 p. 56.
54. Koonz, p. 180.
55. Ibid., p. 5.
56. Ibid., p. 55.
57. Ibid., p. 186.
58. Ibid., p. 196.
59. Ibid., p. 7.
60. Ibid., p. 418.
61. Ibid., p. 420.
62. See Deborah Dwork, discussion of Himmler's plans.
63. Goldhagen, p. 110. Garry Wills, Papal Sin: Structures of Deceit, New York: Doubleday, 2000, and James Carroll, Constantine's Sword: The Church and the Jews: A History, Boston: Houghton Mifflin, 2001, discuss the role of the Catholic Church in forming and abetting anti-Semitism.
64. Goldhagen, pp. 110–111.
65. Ibid., p. 106.
66. Koonz, p. 16.
67. Doris Bergen, Twisted Cross: The German Christian Movement in the Third Reich, Chapel Hill: The University of North Carolina Press, 1996, p. 159.
68. Bergen, p. 228.
69. Mein Kampf, p. 427, and also in Änne Bäumer-Schleinkofer, Nazi Biology and Schools (translated by Neil Beckhaus), Frankfurt am Main: New York: Peter Lang, 1995, p. 3.

70. Mein Kampf, p. 408.
71. Bäumer-Schleinkofer, pp. 4–5.
72. Ibid., p. 63.
73. Ibid., p. 143.
74. Ibid., p. 167.
75. Ibid., p. 143.
76. Ibid., p. 170.
77. Ibid., p. 233.
78. Mein Kampf, p. 421.
79. Ibid., p. 426.
80. Dawidowicz, p. 250.
81. Ibid., p. 253.
82. *The New York Times Magazine,* February 13, 2000; Peter Schneider, pp. 52–57.
83. Kristen Monroe, The Heart of Altruism: Perceptions of a Common Humanity. Princeton: Princeton University Press, 1996, pp. 114–117. The fact that they could not do otherwise does not mean that they lost their will. Rather it means they found the good compelling.
84. Philip Hallie, Lest Innocent Blood Be Shed: The Story of the Village of Le Chambon and How Goodness Happened There, New York: Harper and Row, 1979, p. xii.
85. Hallie, p. 10.
86. Kim Coulter, Hospitality: A Gift of Place, MA Thesis, University of Wisconsin, Madison, 2001.
87. Hallie, p. 10.
88. Ibid., pp. 127–128.
89. Ibid., p. 126.
90. Sheila Fitzpatrick, The Russian Revolution, New York: Oxford University Press, 1982, p. 76.
91. Ibid., pp. 121–122.
92. Robert Tucker, Stalin in Power: The Revolution from Above, 1928–1941, New York: W.W. Norton, 1990, p. 27.
93. John Hazard, The Soviet System of Government, Chicago: University of Chicago Press, 1968.
94. Robert Tucker, p. 442.
95. Robert Argenbright, "Remaking Moscow: New Places, New Selves," *Geographical Review,* 89, 1999, pp. 1–22, pp. 5–6.
96. Ibid., p. 5.
97. Ibid., p. 7.
98. Claude Meillassoux, The Anthropology of Slavery: The Womb of Iron and Gold (translated by Aide Dasnois), London: The Athlone Press, 1991.
99. Kenneth Stampp, The Peculiar Institution: Slavery in the Ante-bellum South, New York: Alfred Knopf, 1956, p. 206.
100. Ibid., p. 208.
101. Ibid., p. 211.
102. Ibid., pp. 211–212.
103. Ibid., pp. 211–212.
104. Clement Eaton, Freedom of Thought in the Old South, New York: Peter Smith, 1951, p. 128.
105. Eaton, p. 199.
106. Ibid., pp. 208–209.
107. Ibid., pp. 213–214.
108. Alberto Manguel, A History of Reading, London: Harper Collins, 1996.

Chapter 5

1. A correspondence theory is likely to hold that truth is ultimately singular, though with infinite manifestations. The same with the good.
2. For an argument that progress will reach a ceiling, see for example, Gunther Stent, Paradoxes of Progress, San Francisco: W. H. Freeman, 1978.
3. Kant expresses this several ways. See Richard Norman, The Moral Philosophers: An Introduction to Ethics, New York: Oxford University Press, 1983.
4. Here I am elaborating on ideas presented in a talk by Tad Mutersbaugh on organic coffee growing, presented in 2001 to the Geography Department at the University of Wisconsin, Madison.
5. The example is from Yi-Fu Tuan, private correspondence.
6. The idea that evil is a contraction of awareness and variety and complexity corresponds to the crosscultural definition of evil in simple societies as a lack or incompleteness. See David Parkin, ed. The Anthropology of Evil, Oxford: Basel Blackwell, 1985.
7. Here I am using the idea of tyranny developed by Walzer in Spheres of Justice.
8. It can be argued that justice is not an end but a means to attain the other virtues.
9. B. Jowett, The Works of Plato, New York: The Tudor Publishing Co., 1937, p. 269.
10. These are from the W. R. M. Lamb translation in Plato: The Loeb Classical Library.
11. Fergus Kerr, Immortal Longings: Versions of Transcending Humanity, Notre Dame, IN: University of Notre Dame Press, 1997, p. 69.
12. Yi-Fu Tuan, Morality and Imagination, Madison: University of Wisconsin Press, 1989, p. 180.
13. See, for example, Fritz Redlich, Hitler: Diagnosis of a Destructive Prophet, Oxford: Oxford University Press, 1998, who sums up the psychiatric consensus that Hitler was not insane but who says he did it willingly. I would part company and say he did not really know what he was doing. Here I am doing the same thing that Kant did when he backed off from the possibility of radical evil and went instead with a weakness of will.
14. Jon Elster, ed., The Multiple Self, Cambridge: Cambridge University Press, 1985, p. 6.
15. After Donald Davidson's weakness of warrant in Elster, ibid., p. 80.
16. James M. McPherson, "Southern Comfort," New York Review of Books, April 12, 2001, p. 28.
17. Hannah Arendt, Eichmann in Jerusalem: A Report on the Banality of Evil, New York: Viking, 1963.
18. See Thomas Nagel, The Possibility of Altruism, Oxford: Clarendon Press, 1970, for a philosophical argument; Kristen Monroe, the Heart of Altruism: Perceptions of a Common Humanity, Princeton: Princeton University Press, 1996, for the altruistic motivation in general; and Jane Mansbridge, ed. Beyond Self-Interest, Chicago: University of Chicago Press, 1990, and Avner Ben-Ner and Louis Putterman, eds. Economics, Values, and Organization, Cambridge: Cambridge University Press, 1998, for the need for altruism in economics.
19. For a postmodern view see Robert Bernasconi, "What Goes Around Comes Around: Derrida and Levinas on the Economy of the Gift and the Gift of Geneology" in Alan D. Schrift, ed. The Logic of the Gift: Toward an Ethic of Generosity, New York: Routledge, 1997, pp. 256–273; and for the theological implications see John Caputo and Michael Scanlon, eds. God, the Gift, and Postmodernism, Bloomington: Indiana University Press, 1999.
20. Yi-Fu Tuan, Dominance and Affection: The Making of Pets, New Haven: Yale University Press, 1984.

Chapter 6

1. Peter Wilson, The Domestication of the Human Species, New Haven: Yale University Press, 1988.
2. Note that Wilson's emphasis on independence of the predomesticated group is not the same as a consciousness about self.
3. Yi-Fu Tuan, Segmented Worlds and Self, Minneapolis: University of Minnesota Press, 1982.
4. Yi-Fu Tuan, Escapism, Baltimore: The Johns Hopkins University Press, 1998. Tuan uses the term not only to mean escaping from reality, but also to mean the good.
5. As I mentioned in the first chapter—note 1—the problematic is drawn from Tuan's formulation on escapism. "A human being is an animal who is congenitally indisposed to accept reality as it is. Humans not only submit and adapt, as all animals do; they transform in accordance with a preconceived plan. That is, before transforming, they do something extraordinary, namely, "see" what is not there. Seeing what is not there lies at the foundation of all human culture." Yi-Fu Tuan, Escapism, p. 6.
6. Susanne Langer, Mind: An Essay on Human Feeling, Vol. II, Baltimore: The Johns Hopkins University Press, 1972, discusses how the ability to conceive of death is a test of the capacity to think symbolically, and that animals do not seem to possess this.
7. The term moral drift is used by Laurence Mordekhai Thomas, Vessels of Evil: American Slavery and the Holocaust, Philadelphia: Temple University Press, 1993.
8. An argument for decency—not humiliating others—as the moral position that may be the most we can expect is Avishai Margalit, The Decent Society, Cambridge: Harvard University Press, 1996.
9. Thomas, op. cit, p. 52.
10. Thomas, p. 87.
11. This is discussed by Thomas, op. cit.
12. Donald Davidson, "Deception and Division" in Jon Elster, ed. The Multiple Self, Cambridge: Cambridge University Press, 1985, p. 80, where he would refer to it as a weakness of warrant. I take this to be virtually the same thing as self-deception.
13. Joseph Ellis, American Sphinx: The Character of Thomas Jefferson, New York: Vintage Books, 1996, p. 19.
14. Ellis, p. 172.
15. Ibid., p. 102.
16. Ibid., p. 102.
17. Ibid., p. 175.
18. Ibid., p. 177–179.
19. Ibid., p. 179.
20. Ibid., p. 175.
21. Ibid., p. 106.
22. George Frederickson, The Black Image in the White Mind: The Debate on Afro-American Character and Destiny, 1817–1914, New York: Harper and Row, 1971, p. 1.
23. Frederickson, The Black Image in the White Mind, p. 1; see also Winthrop D. Jordan, White over Black: American Attitudes Toward the Negro, 1550–1812, Chapel Hill: University of North Carolina Press, 1968.
24. Willard Randall, Thomas Jefferson: A Life, New York: Henry Holt, 1993, p. 303.
25. Jordan, White over Black, pp. 451–452 and Work of Jefferson VI 311. (ed.) Ford. 1791, Phila., Aug. 30, 1791.
26. Jordan, p. 452.
27. But this may be too harsh, for he also considered intellectual expression in other realms as well, though here he was not as impressed, saying for instance, that black

poets appealed more to the heart than the head, although later in his life he admitted that his own poetic sensibilities were not terribly refined. Jordan, White over Black, pp. 437–438.

28. Clement Eaton, Freedom of Thought in the Old South, New York: Peter Smith, 1951.
29. Ibid., p. 19.
30. Eaton, pp. 35–44.
31. Eaton, p. 48.
32. Ibid., pp. 62–63.
33. Eugene Genovese, The Political Economy of Slavery: Studies in the Economy and Society of the Slave South, New York: Vintage, 1965, p. 8.
34. Ibid., p. 31.
35. Thomas, p. 87 and p. 91.
36. Eric Johnson, Nazi Terror: The Gestapo, Jews and Ordinary Germans, New York: Basic Books, 1999, p. 48.
37. Hannah Arendt, Eichmann in Jerusalem: A Report on the Banality of Evil, New York: Viking, 1963, p. 52.
38. The best known of these arguments is Goldhagen's *Hitler's Willing Executioners.*
39. They did not make public announcements about what is real and true (as did the abolutionists), but they did act justly.
40. Tzvetan Todorov, Facing the Extreme: Moral Life in the Concentration Camps, New York: Metropolitan Books, 1996, p. 153.
41. Arendt, p. 81.
42. Michael Burleigh and Wolfgang Wippermann, The Racial State: Germany 1933–1945, Cambridge: Cambridge University Press, 1991, p. 100, and George Mosse, The Crisis of German Ideology; and for those who served in Auschwitz, duty was "an unpleasant assignment that kept them from [the] far more unpleasant prospect of combat on the Russian front." (Kingdom of Auschwitz, 46–47).
43. Todorov, pp. 122–123.
44. Todorov, p. 125. What does one say about Uganda, where thousands of people were macheted to death?
45. Arendt, p. 105.
46. Todorov, pp. 158–159.
47. Ibid., p. 164.
48. Primo Levi, in the Introduction to Höss, p. 7.
49. Arendt, pp. 85–86.
50. Todorov, p. 161.
51. Arendt, pp. 85–86.
52. Höss, p. 49.
53. Ibid., p. 84.
54. Ibid., p. 154.
55. Ibid., p. 96.
56. Ibid., pp. 58–59.
57. Ibid., p. 162.
58. Ibid., p. 163.
59. Ibid., p. 163.
60. "Today I realize that the extermination of the Jews was wrong, absolutely wrong. It was exactly because of this mass extermination that Germany earned itself the hatred of the entire world. The cause of anti-Semitism was not served by this act at all, just the opposite. The Jews have come much closer to their final goal." Höss, p. 183.
61. Höss, p. 183.
62. Arendt, p. 32.
63. Todorov, p. 162.

64. Arendt, p. 116.
65. Ibid., p. 175.
66. Todorov, p. 151.
67. Arendt, pp. 287–288.
68. Todorov, p. 151.
69. Ibid., pp. 158–159.
70. Ibid., p. 402.
71. Ibid., p. 134.
72. Ibid., p. 64.
73. Ibid., p. 281.
74. Ibid., p. 305.
75. John Lukacs, The Hitler of History, New York: Alfred Knopf, 1997, p. 126.
76. Ibid., p. 337.
77. Ibid., p. 351.
78. Ibid., p. 71.
79. Ibid., p. 71.
80. Ibid., p. 71.
81. George Mosse, The Nationalization of the Masses: Political Symbolism and Mass Movements in Germany from the Napoleonic War through the Third Reich, New York: H. Fertig, 1975, p. 80.
82. See Ron Rosenbaum, Explaining Hitler: The Search for the Origins of His Evil, New York: Random House, 1998, p. xxii.
83. Lukacs, pp. 193–194.
84. Lang argues he did it willingly and laughingly or gleefully.
85. As, for example, in the extremely important clandestine publications of World War II, the underground circulation of critical novels and plays in the Soviet Union, and the attempts of slaves to teach each other to read.
86. Paul Adams, Theatrical Territoriality: A Geographical Enquiry into Protest Occupations and Mass Communications, Ph.D. Dissertation, University of Wisconsin Madison, 1993.
87. Elaine Scarry, On Beauty and Being Just, Princeton: Princeton University Press, 1999, p. 25.
88. Yi-Fu Tuan, Passing Strange and Wonderful: Aesthetics, Nature and Culture, New York: Island Press, 1993.
89. Scarry, p. 1.
90. Ibid., p. 5.
91. Fairness is treating them "just-so." Indeed, Rawls discusses exactly how justice is fairness.
92. Scarry, p. 81.
93. Ibid., p. 51.
94. Ibid., pp. 30–31.
95. Ibid., p. 31.
96. Ibid., pp. 52–53.
97. Ibid., p. 214.

Chapter 7

1. Terry Eagleton, "For the Hell of It," London: Review of Books, Feb 22, 2001, pp. 30–32 asks the question whether democracy has only instrumental value or if it also has intrinsic value, where intrinsic concerns the role of democracy in self-determination. My use of intrinsic democracy may be more radical. Not only does it include self-determination, but the role of democracy in helping others.

2. Michael Sandel, "The Procedural Republic and the Unencumbered Self" pp. 12–28 in Shlomo Avineri and Avner de-Shalit, eds. Communitarianism and Individualism, New York: Oxford, 1992, p. 13.

3. Ibid, p. 13. John Rawls, A Theory of Justice, Cambridge: Harvard University Press, 1971, also discusses obligations and duties, both positive and negative (p. 113ff and p. 335ff).

4. The terms he uses change over time, but see, for example, Jürgen Habermas, Moral Consciousness and Communitive Action, Cambridge, MA: MIT Press, 1990, pp. 201–202, and The Theory of Communicative Action, Vol. 1, Boston: Beacon Press, 1981, pp. 319–328.

5. Moral Consciousness and Communicative Action, p. 89.

6. Jürgen Habermas, Communication and the Evolution of Society, Boston: Beacon Press, 1976, p. 186.

7. John Dewey, Democracy and Education, New York: The Free Press, 1944, p. 87.

8. Ibid, pp. 87–88; and John Dewey, Democracy and Education, New York: The Free Press, 1916, p. 212.

9. Dewey, Democracy and Education, 1944, p. 212.

10. See, for example, Chantal Mouffe, "Post Marxism: Democracy and Identity," Society and Space, Vol. 13, 1995, pp. 259–265; Nancy Fraser, "Rethinking the Pulbic Sphere: A Contribution to the Critique of Actually Existing Democracy" Social Text, Vol. 8, 1990, pp. 56–80; and Bonnie Honig, "Difference, Dilemmas, and the Politics of Home" in Seyla Benhabib, ed. Democracy and Difference: Contesting Boundaries of the Political, Princeton: Princeton University Press, 1996, pp. 257–277. For an analysis of radical democracy, see J. Nicholas Entrikin, "Political Community, Identity and Cosmopolitan Place," International Sociology, 14(3), Sept. 1999, pp. 269–282; and "Democratic Place-Making and Multiculturalism," Geografiska Annaler, 84b, 2002, pp. 19–25.

11. Rosalyn Deutsche, Evictions: Art and Spatial Politics, Cambridge, MA: MIT Press, 1998.

12. Nancy Fraser, op. cit., and Bonnie Honig, op. cit.

13. Claude Lefort, Democracy and Political Theory, Minneapolis: University of Minnesota Press, 1988, p. 41.

14. Maurizio Passerin d'Entreves, "Hannah Arendt and the Idea of Citizenship" in Chantal Mouffe, ed., Dimensions of Radical Democracy: Pluralism, Citizenship, Community, London: Verso, 1992, pp. 145–167.

15. Ibid., p. 147. Maurizio Passerin d'Entreves quoting Arendt.

16. Ibid., p. 152.

17. Ibid., p. 153.

18. Chantal Mouffe's emphasis on agonistic instead of antagonistic does not go this far, for it leaves the impression only of argumentation, and not one of changing and expanding one's view.

19. This is a contradiction, according to Kim Coulter's view on hospitality.

20. See Joseph Carens, "Aliens and Citizens: The Case for Open Borders," Review of Politics, 49, 1987, pp. 251–272, and Habermas, Between Facts and Norms, pp. 512–514.

21. Yet, as Serge-Christophe Kolm, "Altruism and Efficiency," Ethics, 94, 1983, pp. 18–65, has pointed out and as we will soon discuss, this was not originally the case, nor does it need now to be.

22. The classic expression of self-interest supposedly leading to the good of everyone is Bernard Mandeville, The Fable of the Bees: Private Vices, Public Benefits.

23. Consider Jane J. Mansbridge, ed. Beyond Self-Interest, Chicago: University of Chicago Press, 1990, and Avner Ben-Ner and Louis Putterman, eds. With a foreword by Amartya Sen, Economics, Values, and Organization, Cambridge: Cambridge University Press, 1998.

24. "In your enjoyment or use of my product I would have the direct enjoyment both of being conscious of having satisfied a human need by my work, that is, of having objectified man's essential nature, and of having thus creating an object corresponding to another man's nature." Comments on James Mill, pp. 227–228, as quoted in Jon Elster, Making Sense of Marx, Cambridge: Cambridge University Press, 1985, p. 454.
25. See, for example, Pranab E. Bardhan and John E. Roemer, Market Socialism: The Current Debate, New York: Oxford University Press, 1993, and especially Roemer's model for example.
26. Robert D. Smither, The Psychology of Work and Human Performance, New York: Longman, 1998, pp. 205–209, for discussion of Maslow.
27. Amartya Sen, Development as Freedom, New York: Alfred Knopf, 1999, p 18.
28. According to Jevons, in Henry Spiegel, The Growth of Economic Thought, Durham, NC: Duke University Press, 1983, p. 521.
29. Albert Hirschman, The Passions and the Interests: Political Arguments for Capitalism Before Its Triumph, Princeton: Princeton University Press, 1977.
30. According to Lewis Hyde, The Gift: Imagination and the Erotic Life of Property, New York: Vintage Books, 1983.
31. Roger Bolton, "Place as Network: Application of Network Theory to Local Communities," 2001 (in press).
32. As in Joseph Carens, Equality, Moral Incentives, and the Market: An Essay in Utopian Politico-Economic Theory, Chicago: University of Chicago Press, 1981.
33. Kolm, op. cit.
34. See again note above on market socialism.
35. My point is not that quantification is to be avoided, but that it may not capture the essence of the problem. As Roger Bolton argues, a rigid resistance to quantifying nonmarket phenomena may actually doom them to inadequate support in political debate, because in that debate many other phenomena are easily described in quantitative terms (many "costs" for example), and thus are privileged. Still, I think we cannot discover the qualities of gift-value without giving them "nonquantitative" room to develop.
36. Personal remark made by Roger Bolton.
37. Hitler, it will be recalled, seemed well aware that gifts must circulate, which is what he wanted to stop when he argued for the reinstatement of what he mistakenly thought to be the former state of science as "occult knowledge." He did not realize that to keep science secret would be to diminish its powers, and even destroy it.
38. Pierre Bourdieu, Homo Academicus, Stanford, CA: Stanford University Press, 1988, conceives of the academic and the pursuit of knowledge as a strategy for personal accumulation of power and prestige.

INDEX